Kulkarni

Robust Process Development and Scientific Molding

Suhas Kulkarni

Robust Process Development and Scientific Molding

Theory and Practice

HANSER

Hanser Publishers, Munich   Hanser Publications, Cincinnati

The Author:
Suhas Kulkarni, 1859 Jamaica Way, Vista, CA - 92081, USA

Distributed in the USA and in Canada by
Hanser Publications
6915 Valley Avenue, Cincinnati, Ohio 45244-3029, USA
Fax: (513) 527-8801
Phone: (513) 527-8977
www.hanserpublications.com

Distributed in all other countries by
Carl Hanser Verlag
Postfach 86 04 20, 81631 München, Germany
Fax: +49 (89) 98 48 09
www.hanser.de

Library of Congress Cataloging-in-Publication Data

Kulkarni, Suhas.
Robust process development and scientific molding : theory and practice / Suhas Kulkarni.
p. cm.
ISBN-13: 978-1-56990-501-2 (hardcover)
ISBN-10: 1-56990-501-0 (hardcover)
1. Injection molding of plastics. 2. Molding (Chemical technology) I. Title.
TP1150.K85 2010
668.4'12--dc22
2010031414

Bibliografische Information Der Deutschen Bibliothek
Die Deutsche Bibliothek verzeichnet diese Publikation in der Deutschen Nationalbibliografie; detaillierte bibliografische Daten sind im Internet über <http://dnb.d-nb.de> abrufbar.

ISBN 978-3-446-42275-9

Production Management: Steffen Jörg
Coverconcept: Marc Müller-Bremer, www.rebranding.de, München
Coverdesign: Stephan Rönigk
Typeset: le-tex publishing services GmbH, Leipzig
Printed and bound by Kösel, Krugzell
Printed in Germany

Dedicated to my Parents
Dr. Mohan P Kulkarni
Jayashree M Kulkarni

Preface

When I interviewed for my second job after I graduated, I was told that if the position was offered to me, I would have to spend my first three days at a seminar on Scientific Molding and Design of Experiments. It was all new to me then. My job was to implement this new technology as a standard across the company. The job was offered to me; I accepted and attended the seminar. Implementing the techniques on the first couple molds was a refreshing change from how I did it before. The scientific method of developing the process left no room for any guess work by applying the theories of polymer science and injection molding. Scientific evidence proved why parts could be or could not be molded consistently within the required specifications. My enthusiasm for the use of these techniques grew as I found more and more evidence of success. Over the next few years, I gave presentations at the local SPE chapter and the attendees wanted to learn more to make their operations efficient. In 2004 I decided to start consulting in the area of Scientific Processing, a term I coined to include all the processes that are involved in the transformation of the pellet to the final product that is shipped out to the customer. My research work on the 'overdrying' of PBT and Nylon was the main driving force to think of the process as being outside of the molding machine and not just what happens in the mold. As my consulting and teaching career expanded, I found many people looking for a resource to learn the basic underlying principles of polymers and plastics and apply them to injection molding. They wanted to understand the why, and then how of Scientific Processing. 'Where can I find this information?' was always a question that was asked. This book is the answer to their question.

Understanding the molding process from the scientific perspective helps in making better decisions to establish the parameters that are involved in controlling the journey of the pellet; from the warehouse to the molding machine and then to its conversion as a molded product. All the parameters are set on the basis of scientific knowledge and experience making the process efficient in terms of productivity. Higher yield, reduced scrap, robust processes, reduced quality inspection, reduced number of process changes leading to less human intervention are some of the benefits of Scientific Processing. This book details the theory and practice of Scientific Processing. There are a lot of 'rules of thumb' in injection molding. My mission is to eliminate them and present a scientific solution. A good example is the size of vents in the mold.

I hope my commitment to researching and understanding of the molding process will continue to give a better insight to the process. I hope to share those with you in the future editions of this book. There are a number of people who are part of the success of writing this book. Some gave me the knowledge, some inspired me to learn more while others gave me unconditional support in this endeavor. It is impossible to thank all of them individually but without all of them this project would not have been accomplished. First and foremost, special mention must be made of my father who introduced me to the fascinating world of chemical research. It is from here that I get my curiosity, creativity and my analytical abilities of problem solving. Thanks to my teachers and professors who not only imparted the knowledge but also instilled in me the value of education through the dedication to their students. It is from here that I get my inspiration to teach and spread my knowledge. Thanks to my family and friends who have supported me and believed in me. It is from them that I get my will power and courage to get past the current frontiers and take a step into an unknown future.

In the production of this book I would like to thank Christine Strohm and the management of Hanser Publications for publishing the book. The sections on cavity pressure sensing and the chapter on rheology were reviewed by Mike Groleau of RJG and John Beaumont of Beaumont Technologies respectively. Thanks to them for their valuable comments. Thanks also to Dave Hart for proofreading the text and making the matter an interesting technical read. Valuable comments from Ravi Khare of Symphony Technologies were included on the DOE chapter. Without the unconditional help of Tim and Violeta Curnutt of Distinctive Plastics I would have not had the chance to experiment with many of the theories and applications put forward in this book. Special thanks to them for letting me make Distinctive Plastics my home during the book writing process. I am often told I am an effective teacher with clear concepts in polymer science and rheology - I have picked the teaching skills and the knowledge from Prof. Basargekar – my sincere acknowledgements to him. Under the leadership of Vishu Shah I conducted a few successful seminars with the Society of Plastics Engineers. These seminars gave me the fuel and material for this book. Thanks to Vishu not only for the opportunities of the seminars but also for being a professional guide and a personal friend. I would also like to acknowledge the efforts of John Bozzelli and Rod Groleau for their pioneering work in Scientific Molding and raising its awareness in the molding community.

To my alma maters, Maharashtra Institute of Technology, Pune, India and University of Massachusetts, Lowell, USA : Hidden in one of your foundations' bricks are the enriching roots to my success. Thank You.

Suhas Kulkarni
FIMMTECH Inc.
Vista, CA.
January 2010

Contents

1 Introduction to Scientific Processing

1.1 The Evolution and Progress of Injection Molding

Injection molding and extrusion are the most common techniques employed in the manufacture of plastic products. Injection molding of plastics began as an idea by the Hyatt brothers for the manufacture of billiard balls. The idea was borrowed based on a patent by John Smith to inject metal castings. Since then, injection molding of plastics has come a long way. The technique became a popular way to fabricate plastic parts because of the simplicity of the concept, efficiency of production, and the possibility of producing intricate parts with fine details.

The art of injection molding evolved to its present state due to a few key reasons. The requirements of the molded parts became more stringent because of the advances in the fields of science and technology. The demand for tighter tolerances and more complex parts increased and is ever increasing. A required tolerance of a couple thousandths of an inch on a one inch dimension is not uncommon these days. Parts requiring innovative designs, especially designed for assembly (DFA) or parts molded from different materials in the same mold (multi-material molding) are now commonplace. As polymer materials were developed for injection molding, the requirements of processing changed. The discovery of the different morphologies of polymers and the need for better melt homogeneity in molding led to the introduction of the injection screw. Various designs for material-specific screws have followed since. The use of high temperature materials that have high melting points and need high mold temperatures have led to the use of high-temperature ceramic heaters and mold temperature controllers providing higher heat capability. Innovations in electrical and electronic technologies paved the road for machines that could be better controlled, accurate, and efficient. Response times for hydraulic valves can be in milliseconds. All electric machines and hybrid machines are gaining popularity because of their consistency and accuracy. The real time processing parameters of a molding machine can now be viewed from any part of the world via an internet connection and therefore machine production can be monitored or machines can be debugged online. All these features are becoming a common practice among manufacturers. Even some auxiliary equipment can now be debugged and programmed by the suppliers via an internet connection. For the machines tied into the company ERP system, automated messages can be sent to the managers and supervisors about the machine status and quality issues. The need for efficiency and the requirements for advanced product features have dictated the need for innovations in injection molding over the years.

1.2 The Molding Process

The actual molding process has been traditionally defined as the inputs to the molding machine. These are the settings of speeds, pressures, temperatures and times such as injection speeds, holding pressure, melt temperature and cooling time. These are inputs one would set at the molding machine and record on a sheet, commonly called the Process Sheet. However, the word process now needs to be redefined as the complete operation that encompasses all the activities the plastic is subjected to inside a molding facility – from when the plastic enters the molding facility as a pellet to when it leaves the facility as a molded part. For example, the storage of the plastic, the control of the drying of the plastic, and the post mold shrinkage of the part can have a significant influence on the quality of the part. During this journey of the pellet, every stage can have a significant effect on the final quality of the part or assembly. Naturally, understanding every stage now becomes imperative if we would like to control the quality of the molded part. Molding a part that meets the quality requirements is not the real challenge. The real challenge is molding parts consistently; cavity to cavity, shot after shot, and from one production run to another meeting all the quality requirements and with the least amount of effort and maximum efficiency.

1.3 The Three Types of Consistencies Required in Injection Molding

The aim of developing a molding process should be to develop robust processes that would not need any process modifications once the processes are set. Process consistency leads to quality consistency, see Fig. 1.1. We look for three different types of consistencies: cavity-to-cavity consistency (Fig. 1.1 a), shot-to-shot consistency (Fig. 1.1 b), and run-to-run consistency (Fig. 1.1 c). Cavity-to-cavity consistency is required in multi-cavity molds so that each cavity is of the same quality level as the other cavities. Shot-to-shot consistency implies that

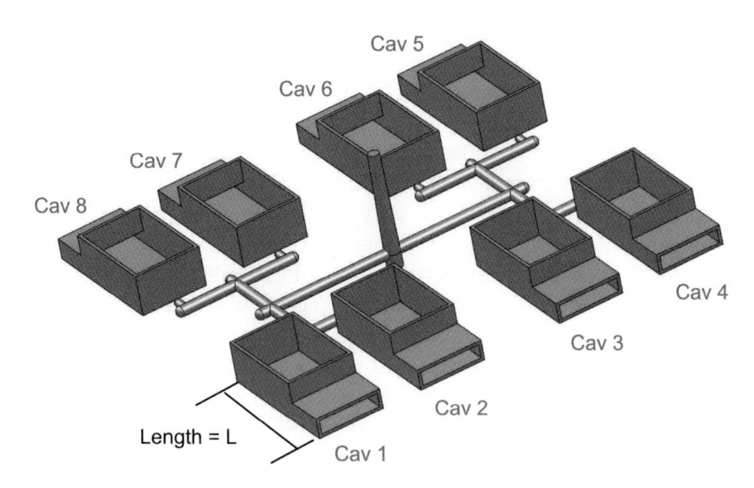

(a) Cavity to cavity consistency

Figure 1.1 The three types of consistencies required in injection molding

every consecutive shot would be identical to the previous shot, or the first shot is identical to the last shot of the production run with the process parameters remaining the same during the entire production run. When the process parameters from two different runs are identical and they produce the same quality parts, then this is called run-to-run consistency. Robust and stable processes always yield consistent quality parts with one established process.

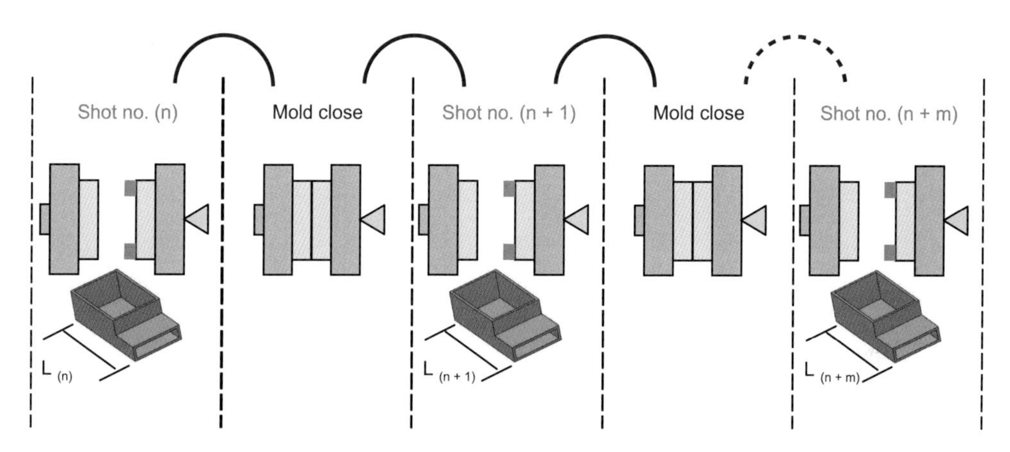

Shot to shot consistency:
Part length, $L_{(n)} = L_{(n+1)} = L_{(N+.....)} = L_{(n+M)}$

(b) Shot to Shot consistency

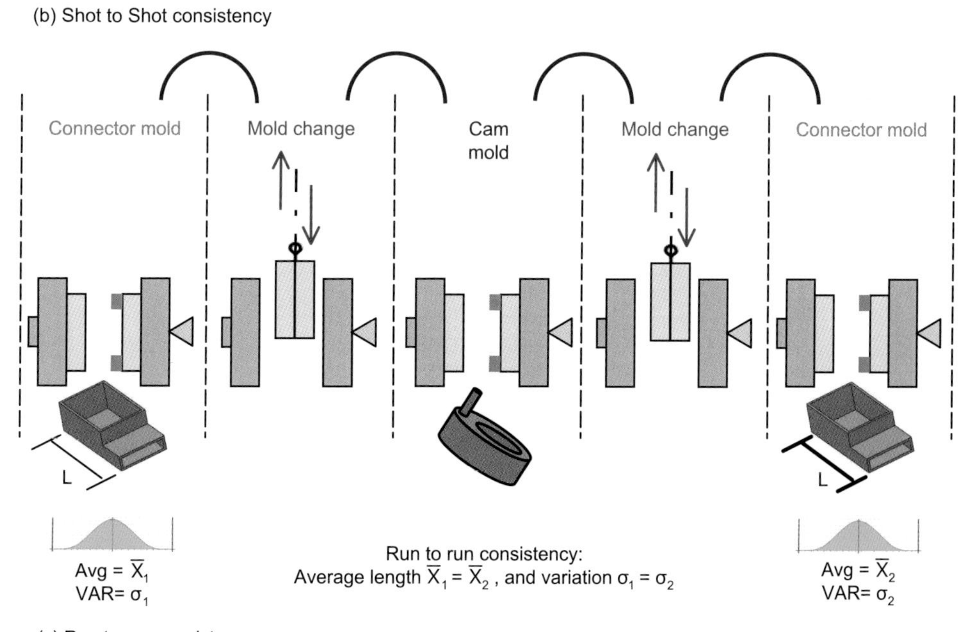

(c) Run to run consistency

Figure 1.1 *(continued)* The three types of consistencies required in injection molding

1.4 Scientific Processing

Scientific Processing is the process of achieving consistency in part quality via the application of the underlying scientific principles that control the parameters of the molding process. To achieve this consistency, we must be able to control every activity that is taking place in the process and to control every activity, we must understand the underlying scientific principles. The goal of scientific processing should be to achieve a robust process. Achieving robustness in each of the stages that the pellet travels through automatically translates to an overall robust process. The term consistency must not be confused with the parts being within the required specifications. A consistent process will produce parts that will reflect the consistency but the parts may be out of specifications. In this case, the mold steel must be adjusted to bring the parts within the required specifications and the process must not be altered.

The term 'Scientific Molding' was coined and promoted by a two pioneers in the field of injection molding, John Bozzelli and Rod Groleau. Their principles are widely used today and are industry standards. Scientific Molding deals with the actual plastic that enters the mold during the molding operation at the molding press. Scientific Processing is the complete process from when the pellet enters the facility and leaves the facility as a finished product. Figure 1.2 shows the journey of the pellet.

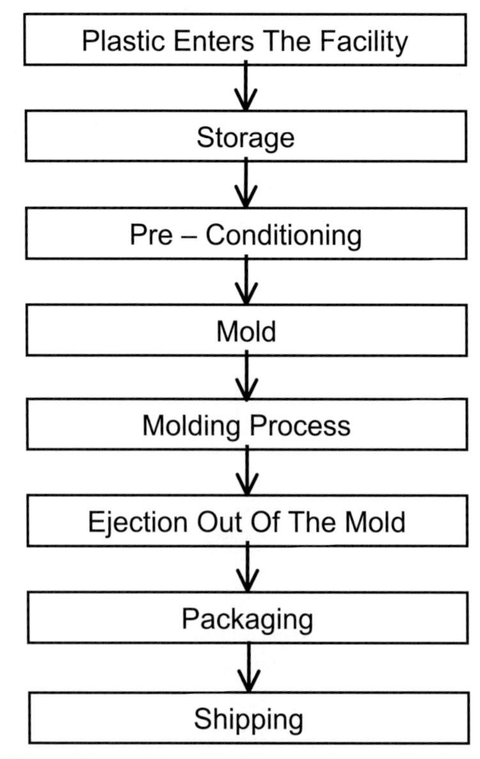

Figure 1.2 The journey of the pellet and the critical factors that need to be controlled

1.5 The Five Critical Factors of Molding

The final molded part is a result of five critical factors that need to be carefully selected as shown in Figure 1.3:

- Part design
- Material selection
- Mold design and construction
- Molding machine
- Molding process

Each of these factors plays a very important role in the production of the molded part and therefore everyone of them has to be optimized for producing the molded part. It is not just the performance of the part but also the consistent molding of the part in production.

1.5.1 Part Design

The concept of the part starts with the engineer designing it. The part must be designed for molding and all the design rules for plastics must be considered. Rules for plastic part design are considerably different than those used for metal part design because of the inherent nature of the plastic. For example, to avoid sink defects in the plastic part, thick sections cannot be present. Additionally, all corners must have a radius to avoid stress concentration and premature failure. With the growing cost of labor and the need for efficiency in the manufacturing process, the part designers now face the added challenges of designing parts for assembly along with those molded parts that utilize multiple materials, commonly referred to as multi-component molding or multi-material molding.

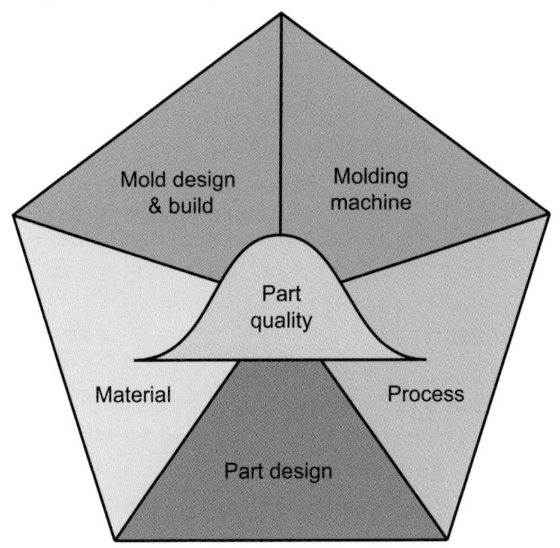

Figure 1.3 The five factors influencing part quality consistency and process robustness

1.5.2 Material Selection

Based on the part design and the part performance requirements, the plastic material must be selected. In addition, the part design may require a special plastic material or a special additive to be added to the base plastic for performance. If a thick section must be present, a filled material may need to be selected or if there is a sliding surface, then an additive reducing the coefficient of friction may need to be added to the plastic. Material selection should typically be done when the basic part design is done. Additional smaller changes can be done concurrently.

1.5.3 Mold Design and Construction

Once the part design and material selection is complete, the mold must be designed and constructed such that it is robust enough to withstand the molding process and the plastic material. For example, during the molding process, the mold can be subjected to high mechanical stresses, especially during the plastic injection and the packing phases. The gates are high-wear areas and there are several places where the air needs to vent out for the plastic to enter the mold. Some plastic materials will require special attention and the mold must be specifically designed with the material in mind. Shrinkage may vary considerably from material to material. All these material specific factors must be considered. The required number of parts over the life of the mold is another factor that will dictate the actual materials of construction. Wear on the mold components must be considered, as the materials chosen to build the injection mold and mold cavities will impact the overall life of the mold and associated amount of maintenance required to keep it production worthy.

1.5.4 Machine Selection

Selecting the right machine for the mold should be done once the mold design is complete. It can be done concurrently during the mold construction stage. The machine plays a very important role in the stability of the molding process. For example, machines with large shot sizes must not be used to mold small shots because the part quality consistency will suffer. Vice versa, using a large percentage of the shot size can give rise to problems with melt homogeneity and therefore issues with fill and dimensions. Small molds must also not be mounted in large machines for fear of mold damage due to excessive clamp tonnage being applied.

1.5.5 Molding Process

Process optimization is the last step before the mold is released into production. This book will cover this topic in detail. If the above four factors and activities are not properly selected or performed, process optimization can be a challenge, if not impossible, without incurring significant cost and delay to the project. At this stage, it is usually very late in the project timeline to make any changes to the part design or mold design, especially because of the cost and time involved. An improperly constructed mold can have a very narrow process window leading to a process that will tend to be unstable. If the material selected is not capable of holding the tolerances, no process will be able to produce satisfactory parts.

1.6 Concurrent Engineering

There are various departments involved in the production of the molded part and therefore regular meetings between the different departments must be held. Each department will have specific knowledge of the selection process and can contribute not just to the process but more importantly predict issues once the mold comes over to their department. For example, getting the process engineer involved in a mold design can help in part orientation in the mold for easy removal, or the mold maker can get help with vent locations based on the process engineer's experience. Involving the quality engineer can help the process engineer understand the required tolerances in the design stage. If the tolerances seem to be unrealistic, they can go back to the product designer for wider tolerances or a material change. There are a lot of benefits associated with implementing concurrent engineering in injection molding. A section is devoted to this topic in this book. In the chapters that follow, the reader will be introduced to the underlying scientific principles to achieve a robust molding process. This understanding will then help in the application of these principles, to develop a robust process and to troubleshoot problems that occur in production. The chapters have been written in a logical sequence to build the readers' knowledge as one would require it or should learn it. However, if the reader is familiar with the topic, he or she can bypass some in favor of other chapters containing the desired information.

Suggested Further Reading

1. Osswald, T.A., Turng, L., Gramann, P.J. (Eds.), Injection Molding Handbook (2007) Hanser, Munich
2. Kulkarni, S.M., Injection Molding Magazine (June 2008) Cannon Publications, Los Angeles, USA

2 Introduction to Polymers and Plastics

The term plastic is most commonly used when referring to injection molding materials. Plastics are a class of long-chain molecules called polymers. When polymers have certain properties they are called plastics. Since most of the commercially molded polymers fall under the classifications of plastics, we shall refer to these materials as plastics in this book. The other most commonly molded polymers are thermoplastic elastomers (TPEs) that have the same molding characteristics as plastics but different properties when molded. When referring to these materials, these will be mentioned as TPEs. To understand the concept of injection molding of plastics, a basic understanding about polymers, their properties and the additives that are incorporated into them is required. This chapter will discuss the topic of polymers and their application to the injection molding process.

2.1 Polymers

Every particle in the universe is composed of atoms. Atoms in turn combine to form molecules. A molecule of water is made up of two atoms of hydrogen and one atom of oxygen. Polymers are very large molecules that have several identical molecules joined together. An ethylene molecule attaches itself to another ethylene molecule and when several thousands or millions of such molecules join with each other, a polyethylene molecule is formed. 'Poly' means many and 'mer' means part. A polymer is many parts chemically joined together. The basic single unit from which it is synthesized is called a monomer, 'mono' meaning one. Polymers are also called macromolecules. The process of converting monomers to polymers is called polymerization. Polymers can also be synthesized from multiple monomers. For example, ABS is synthesized from three different monomers, acrylonitrile, butadiene, and styrene.

Polymers have been around since the beginning of time. DNA, the basic unit of life, is a polymer found in all plants and animals and is a naturally occurring polymer. Today, almost all commercially available polymers are synthesized from natural ingredients. The first commercially synthesized polymers were materials such as ebonite in the late 1800s. Interestingly, the widely used polyolefins gained commercial importance only in the late 1950s, almost couple decades after the introduction of polyvinyl chlorides, nylons, and polyesters. Recently introduced polymers are based on biomaterials and nanotechnology.

Polymers are synthesized from monomers via a chemical process. There are mainly two types of polymerization processes, the *addition polymerization* process and the *condensation polymerization* process. In the addition polymerization process, a catalyst initiates the polymerization reaction and each monomer adds onto the next monomer until all the monomers are polymerized. A common example of an addition polymer is polyethylene. Polyethylene is polymerized from ethylene monomer, which is a gas at room temperature. The double bond in the ethylene molecule breaks and a bond with an adjacent ethylene molecule is formed. The process continues and the result is a large molecule with high molecular weight. The polymerization process is shown in Figure 2.1.

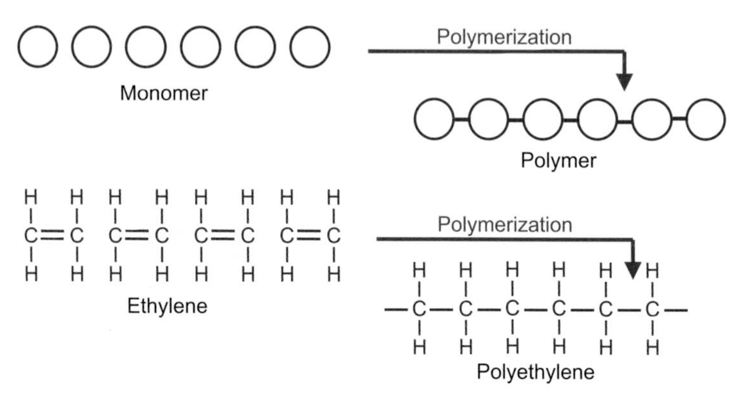

Figure 2.1 The process of polymerization and formation of polyethylene

In condensation polymerization too, each monomer adds on to the next monomer, but this chemical reaction also produces a low molecular weight byproduct that has to be continuously removed out of the system for the polymerization to continue. Condensation polymers are usually polymerized from two or more families of monomers. Nylons and polyesters are examples of condensation polymers. A nylon (chemical name: polyamide) is polymerized from the monomer families of diamines and diacids, as shown in the chemical reaction.

$$nH_2N - R - NH_2 + nHO_2C - R' - CO_2H \rightarrow$$
$$H - (- NH - R - NHCO - R' - CO -)_n - OH + (2n\text{-}1)H_2O$$

R and R' are the characteristic groups that are present in the monomer. In this case, water is the byproduct. Based on these groups, different types of nylons can be produced. The unit in parenthesis repeats itself to form the polymer. If R is $(Ch_2)_6$, then the first monomer is hexamethylene diamine and if R' is $(CH_2)_4$, the second monomer is adipic acid. The polymer that is synthesized from these two monomers is poly(hexamethylene adipamide), commonly called Nylon 6,6.

2.2 Molecular Weight and Molecular Weight Distribution

The repetition of the monomer units causes the molecular weight to increase. The number average molecular weight is the addition of the molecular weights of each of the molecules divided by the number of molecules. Most commercial polymers have a number average molecular weight between 40,000 and 200,000, with some having extremely high numbers. Ultra high molecular weight polyethylene (UHMWPE) is an example for a molecular weight in the range of 1 – 6 million. Molecular weights for greases and soft waxes range between 500 and about 3000, whereas some tough and brittle waxes have molecular weights between 3000 and 10,000 [1]. When the attraction between the molecules (intermolecular forces) is high, the materials can gain sufficient mechanical properties at lower molecular weights. Polyamides and polyesters are examples of polymers with strong intermolecular forces. In materials such as polyethylene, where intermolecular forces are low, high molecular weights are required to achieve acceptable mechanical properties. In general terms, molecular weights for polyethylenes are higher those that for nylons or polyesters. UHMWPE was developed for applications in which polyethylenes were the right choice except for their mechanical properties. In most cases, the mechanical properties reach a plateau with increasing molecular weights. Other properties are also affected by molecular weight. Of particular interest to molders is the viscosity of the polymer where an increase in molecular weight results in the increase in viscosity. For melt processing, a certain minimum viscosity is essential for the formation of a processable and homogeneous melt. Processability increases with molecular weight but due to the increase in viscosity, the energy required to process also increases and reaches a point where the increase is not practical for melt processing, see Fig. 2.2.

The addition of the monomers during the polymerization process (both addition and condensation), is completely random. It is difficult to control the growth of the molecules which results in molecules of various lengths and therefore varying molecular weights. This in turn leads to a distribution of the molecular weight in the polymer called the molecular weight distribution (MWD) of a polymer. The MWD is an important factor in processing. The lower molecular weight units melt faster than the high molecular weight units. In injection molding, the plastic needs to be injected into the mold as fast as possible to make sure the molecules do not freeze off in the cold mold during injection. If this happens, the part will not fill completely and/or will have built up internal stresses when ejected out of the mold. A narrow MWD ensures that all the molecules are molten during approx. the same period of time. When the residence time in the barrel of the injection molding machine reaches the upper limit, the possibility of molecular breakdown or degradation, resulting in the loss of properties in the final product, increases. This is another reason why a low MWD is desired. However, in the case of extrusion, melt strength is an important element of the process. In this case, the higher molecular weight units have higher viscosity and help to carry the molten lower molecular weight units to form the extrudate. Therefore, a broader MWD is pre-

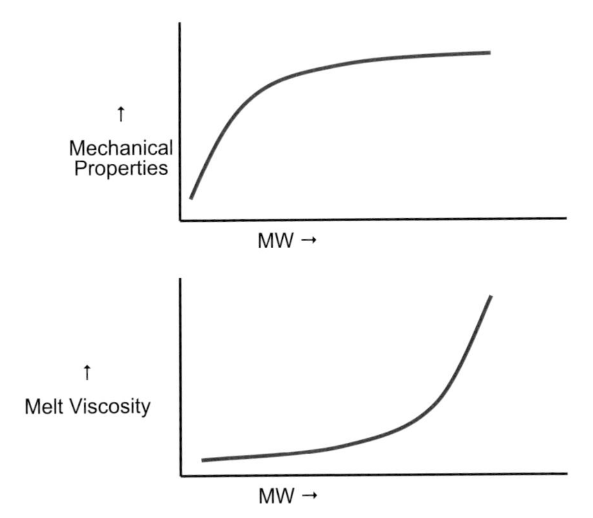

Figure 2.2 Effect of molecular weight on mechanical properties and the viscosity of polymers

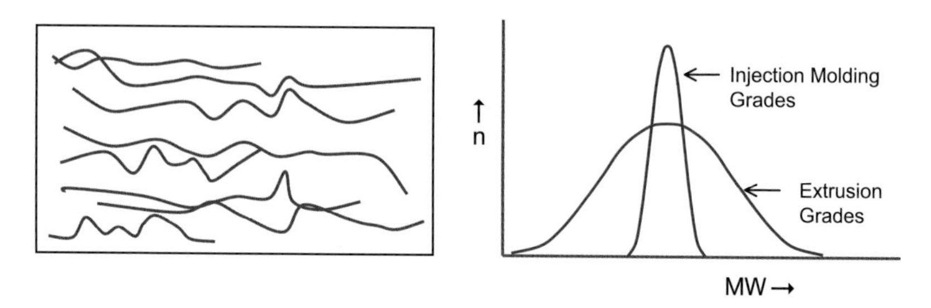

Figure 2.3 Molecular weight distribution for injection molding and extrusion grades

ferred. A narrow MWD would result in the loss of melt strength and therefore in the loss of the shape and characteristics of the extrudate profile. The residence time inside the barrel of an extruder is short because extrusion is a continuous process and therefore the risk of degradation is low. This difference in MWD is the decisive factor whether a resin is an injection molding grade or an extrusion grade. Sometimes, extrusion grades are used in injection molding because the viscosity can be low enough to fill the cavities effectively and consistently. The opposite would be less likely, where injection molding grades are used in extrusion. Figure 2.3 shows the difference in the MWD for injection molding and extrusion grades.

2.3 Polymer Morphology (Crystalline and Amorphous Polymers)

Polymer morphology is the type of arrangement of the molecules in a polymer sample. Based on the different ways the molecules can be arranged, there are two types of polymers – amorphous and crystalline polymers. In amorphous polymers, the molecules are randomly present without any structure or arrangement. Under a high-power microscope, this sample would look like a big bowl of cooked spaghetti. In the case of crystalline polymers, there are certain regions of the sample where the molecules are present in a highly ordered and structured manner. Each of these regions is called a crystallite. No polymer can be completely crystalline and there are always areas where the molecules are present in a random manner. A polymer is therefore truly semi-crystalline in nature, where crystallites are present in the midst of amorphous regions. Part of a molecule can be present in an amorphous region and part of it can be present in a crystallite. The degree of crystallinity refers to the amount of crystallites present in the sample. The difference between the two morphologies is shown in Figure 2.4. Table 2.1 shows the degree of crystallinity for some common polymers.

There are two main reasons why a polymer can be amorphous or crystalline – the geometric regularity of the polymers and the strength of the intermolecular forces between them. Geometric regularity is the arrangement of the groups on the main chain. Stereoregular polymers are those where the monomer segments on the main chain are in the same regular configuration. This regularity helps the chains to pack closer and easier together, just as the

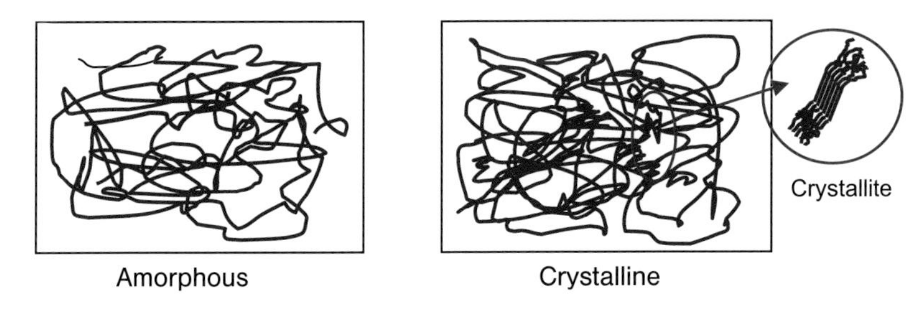

Figure 2.4 Arrangement of molecules in amorphous and (semi)crystalline polymers

blocks in the game of Tetris. In the case of atactic polymers, where the monomer segments are randomly oriented, the chance for packing of the molecules is reduced, leading to an amorphous polymer. During the polymerization process of some polymers, it is possible to control the orientation of the monomers and produce stereoregular polymers. Polystyrene is an example of such a controlled polymerization. Regular polystyrene is atactic and therefore amorphous. However, it can be polymerized by using certain metallocene catalysts to form a stereoregular polymer that makes the polystyrene semi-crystalline. The properties of crystalline polystyrene are far superior and their mechanical and chemical properties excel. The molecular structures of atactic and syndiotactic polystyrene are shown in Figure 2.5.

Another common example is linear polyethylene, which is highly crystalline and branched polyethylene, which is amorphous. Intermolecular forces also play a role in determining crystallinity. The stronger the attraction between the groups of molecules, the higher is the crystallinity. Nylon is an example where the intermolecular forces are high, see Figure 2.6. The hydrogen atom of one molecule has a strong affinity to the oxygen atom of the adjacent molecule, causing the chains to get closer and pack better. This effect can also be observed in the case of polyesters.

Table 2.2 lists the morphologies of some common polymers. However, it must be noted that morphologies can be altered during the manufacture and during the melt processing of the polymer. For example, mold temperature plays a very important role in the development of crystallinity. Low mold temperatures do not favor the development of crystallites. This is described later in the chapter.

Table 2.1 Degree of Crystallinity for Common Polymers [2]

Polymer	Degree of crystallinity
High density polyethylene	0.80
Isotactic polypropylene	0.63
Poly(ethylene terephthalate)	0.50
Nylon 66	0.70
Nylon 6	0.50

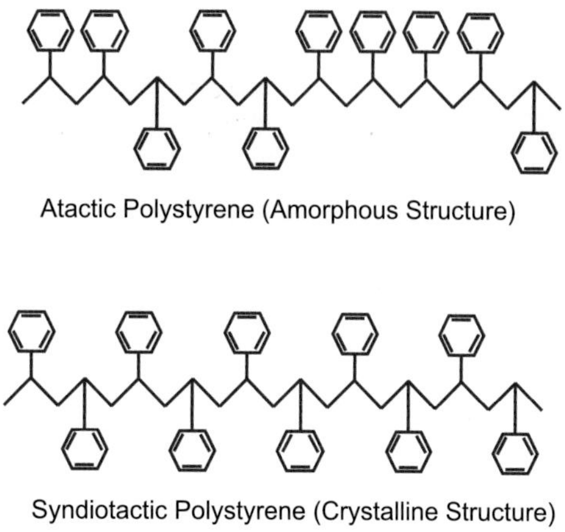

Atactic Polystyrene (Amorphous Structure)

Syndiotactic Polystyrene (Crystalline Structure)

Figure 2.5 Atactic and syndiotactic polystyrene

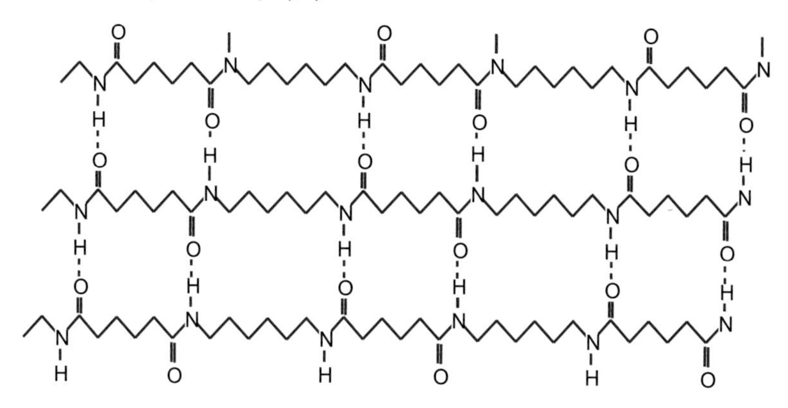

Figure 2.6 Intermolecular attraction resulting in an increase in polymer crystallinity

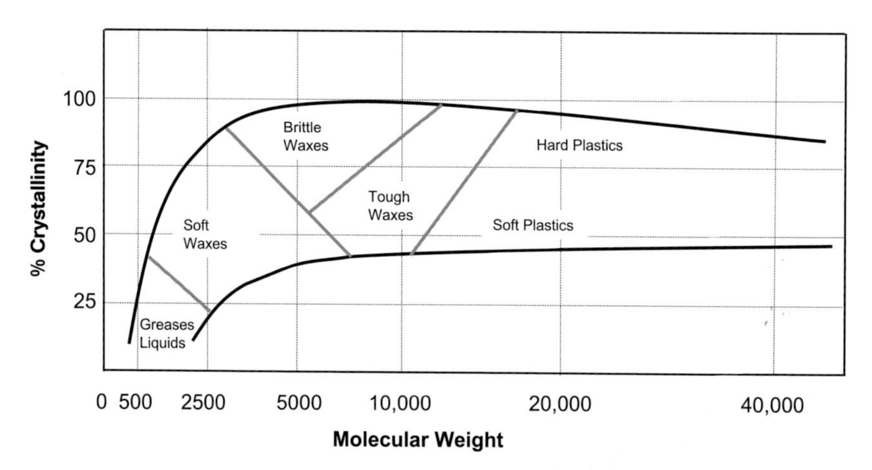

Figure 2.7 Effect of molecular weight and crystallinity on the mechanical properties of polyethylene [1]

The relationship between the percentage crystallinity and molecular weight defines the use of the polymer, see Figure 2.7 that defines the use of polyethylene based on this relationship. Low molecular weight and low crystallinity polyethylene is typically used more in soft waxes and greases, whereas high molecular weight and highly crystalline materials tend to be hard plastics.

Table 2.2 List of Amorphous and Crystalline Polymers

Polymer	Chemical name	Amorphous	Semi-crystalline
ABS	Acrylonitrile butadiene styrene	Y	
ASA	Acrylonitrile styrene acrylate	Y	
GPPS	General purpose polystyrene	Y	
HDPE	High-density polyethylene		Y
HIPS	High-impact polystyrene	Y	
LCP	Liquid crystalline polymers		Y
LDPE	Low-density polyethylene		Y
PA	Polyamide (nylons)		Y
PAI	Polyamide imides	Y	
PBT	Polybutylene terephthalate		Y
PC	Polycarbonate	Y	
PEEK	Polyether ether ketone		Y
PET	Polybutylene terephthalate		Y
POM	Polyoxymethylene (acetal)		Y
PP	Polypropylene		Y
PPS	Polyphenylene sulphide		Y
PSU	Polysulphone	Y	
PVC	Polyvinyl chloride	Y	
SAN	Styrene acrylonitrile	Y	

2.4 Role of Morphology in Injection Molding

The molding characteristics of crystalline and amorphous polymers are different. Crystallites are formed because of high molecular attraction and because of the possibility of the chains being unhindered to form the bond. Sometimes just the presence of another molecule or a side chain prevents crystallization. For melt processing, the crystallites must be dissolved and the chains separated from each other, in order to reduce the viscosity and inject the melt into the mold. It is this basic nature of forming and dissolving of the crystallites that dictates various differences in processing and melt behavior.

2.4.1 Differences in Shrinkage Between Amorphous and Crystalline Materials

Shrinkage is the volumetric change between the melt phase and the glassy or rubbery phase. As the temperature increases, the molecules gain more and more energy, become mobile, and move away from each other. This results in an increase in the volume of the polymer. The intermolecular volume is called free volume. As the polymer cools, the opposite takes place and the free volume reduces, a process known as shrinkage. In crystalline polymers, the movement of the molecules away from each other is much greater compared to amorphous polymers. As the molecules cool, they settle back into a highly structured and closely packed array, which is another contributing reason for their high shrinkage. The absence of such a structure in amorphous polymers negates the need for the molecules to find a definite resting place during the cooling process, resulting in a lower shrinkage value compared to crystalline plastics. Shrinkage values for ABS, an amorphous polymer, are approx. 0.5 – 0.8 % compared to some nylons or acetals that can exhibit shrinkage values of up to 2.5 %.

2.4.2 Melt Processing Range

Similar to an ice crystal, the polymer crystallite needs a specific and definite amount of energy to melt. This is called the crystalline melting point and at this particular temperature the crystallite melts. In crystalline plastics, melting occurs over a fairly narrow range of temperatures; this melting range typically covers approx. 20 °C (30 to 35 °F), see Figure 2.8. For example, PBT (Valox 420 from Sabic Innovative Plastics) needs to be processed between 248 and 265 °C (480 to 510 °F), i.e., within a range of 18 °C. The temperature is not as specific and sharp as in the case of simple molecules because of the presence of the amorphous regions and other attraction forces. The lower the percentage of crystallinity, the broader is the processing range.

In the case of amorphous plastics, there is no melting point but the molecules soften over a range of temperatures. There is a recommended processing temperature range for amorphous plastics. In this range, the viscosity of the plastic is low enough to flow and fill the cavities. For example, ABS (Cycolac from Sabic Innovative Plastics) can be processed from 218 to 260 °C (425 to 500 °F), a range of 42 °C.

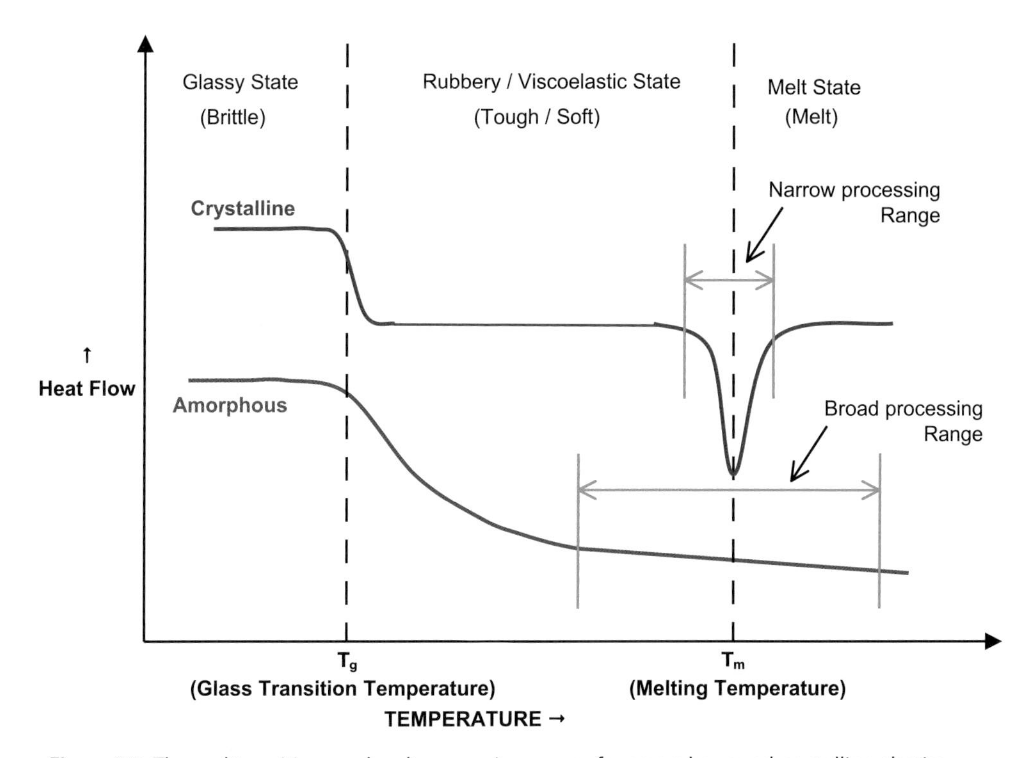

Figure 2.8 Thermal transitions and melt processing ranges for amorphous and crystalline plastics

2.4.3 Mold Filling Speed

Viscosity of the plastic and melt temperature are inversely related. As the plastic temperature increases, the viscosity decreases. As the plastic flows through the cold mold, the temperature of the plastic drops and the viscosity increases. Since for crystalline plastics, the processing range is narrow, the temperature of the flow front must always be higher than the minimum required melt temperature. In the above example of the PBT, the flow front temperature must never drop below 248 °C (480 °F) before the cavity is completely filled. This narrow processing range of crystalline polymers therefore dictates the fact that the plastic must be injected as fast as possible into the mold.

In the case of amorphous plastics, because the plastic stays viscous over a broader range of temperatures, slow injection speeds are permissible, as long as the flow front stays above the minimum processing temperature. This is typical in the molding of lenses and other optical parts.

2.4.4 Mold Temperatures

Similar to the melting of the crystallites at a particular temperature, the crystallites also start to form at a particular temperature, a temperature lower than the melt temperature. This is called the crystallization temperature. This temperature supplies the necessary energy for the formation of the crystallites. If the mold is too cold, the polymer will not be able to receive this energy, preventing the formation of the crystallites. This leads to the loss of the properties in the molded part. Material suppliers conduct extensive research to determine suggested operating mold temperatures. Therefore, it is highly recommended to stay within the recommended specifications for crystalline materials. This concept is further explained in Section 2.5 on thermal transitions.

Since crystallites are absent in amorphous plastics, the mold temperature range can be broader and can be especially extended towards the lower temperatures. Because crystallites do not need to be formed, the molecules are not looking for any particular amount of energy; therefore, lower temperatures are acceptable. However, care has to be taken to prevent molded-in stresses when parts are produced using extremely cold molds.

2.4.5 Barrel Heat Profile

In the injection molding barrel, the screw performs the function of conveying and melting the polymer. The base of the screw is where the polymer pellets first come in contact with the screw. This section is designed to convey and then soften the pellets. In case of crystalline polymers, the crystallites need a lot of energy to melt, so this section, which is usually the second heating zone from the back of the screw, is set at a higher temperature than the next zones in order to initiate the softening of the molecules. But because crystalline plastics can also be heat sensitive or cannot stand high temperatures for long periods of time, the temperature in next heating zone is reduced. This leads to a heating profile that has a hump in the middle, which is typical for crystalline plastics. In case of amorphous plastics, such a profile is not necessary because they need less energy to soften and can tolerate longer residence times in a heated barrel, as is shown in Figure 2.9.

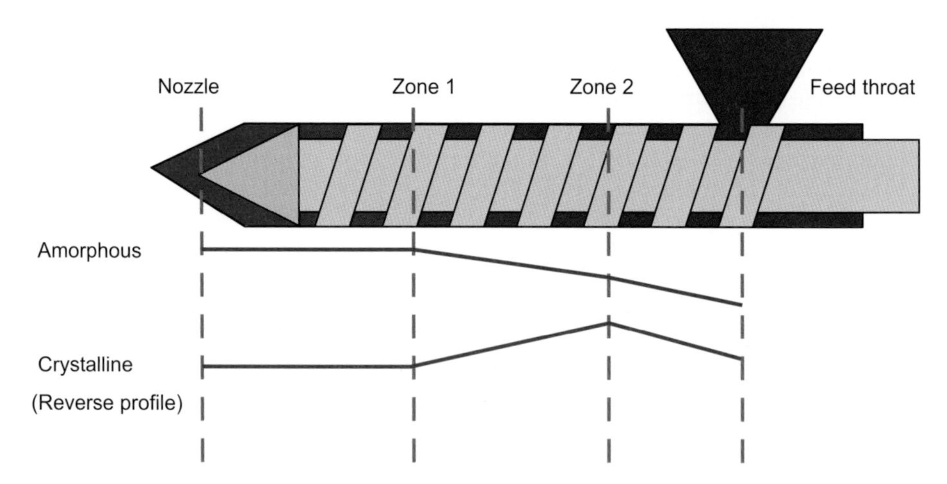

Figure 2.9 Barrel heat profiles for amorphous and crystalline plastics

2.4.6 Screw Recovery Speeds

The barrel heaters provide heat on the outside of the cylinder. Because plastics are bad conductors of heat, the plastic closer to the screw and furthest from the wall of the barrel requires additional heat to become plasticized. This additional effective melting energy comes from the shear friction caused by the rotating screw. High screw speeds generate high amounts of shear, helping the crystallites melt and ensuring melt homogeneity. For amorphous plastics, due to the low energy required to melt the plastic, high screw speeds are not critical. In fact, high screw speeds can degrade the material and cause defects such as splay.

2.4.7 Nozzle Temperature Control

Nozzle temperature control is critical while processing crystalline polymers. Not only must the temperature of the nozzle be maintained within the processing range, it should be maintained within a temperature range that is much narrower than the processing range. This is especially important during the static phases of the plastic flow, which follows the holding phase of one cycle and precedes the injection phase of the next cycle. Lower temperatures tend to freeze off the plastic in the nozzle tip making it impossible to inject the next shot. At higher temperatures, all the crystallites will have disappeared making the viscosity very low and causing what is commonly called nozzle drooling. There are some innovative designs for nozzle tips available in the market to help prevent this problem. With amorphous plastics, the broad processing range helps maintain the viscosity in the nozzle and prevents freeze off or drool.

2.4.8 Cooling Times

Once the crystallites are formed, the molded part gains sufficient strength and only needs a small drop in temperature to be ejected out of the mold. Therefore, in the case of crystalline materials, the cooling times are shorter compared to amorphous plastics (for the same part thickness). Nucleating agents are added to some crystalline plastics to accelerate crystallization, thus reducing the cooling time even further. Nucleating agents have no effect on amorphous materials because there are no crystallites to be formed.

2.4.9 Mechanical Properties

The crystallites provide mechanical strength to the polymer. They are like a rope, rather than a bundle of grass, providing strength. Generally speaking, crystalline plastics have higher mechanical properties than amorphous plastics. Therefore, most are referred to as engineering resins. However, with the advent of new technologies and the discovery of new additives, the properties of amorphous materials can be easily modified to match those of crystalline materials.

2.4.10 Optical Clarity

Most amorphous polymers in their natural and unmodified state are optically clear. The distance between the molecules is large, allowing the wavelengths of light to pass through and making them transparent – polystyrene is an example of this. For crystalline polymers, the packing of the molecules does not allow the passage of light and therefore they are usually opaque. As the degree of crystallinity decreases, the materials tend be translucent. As will be discussed in the next section, melts of any polymer, crystalline or amorphous, are always amorphous and therefore a purge of an unfilled melt will always look clear. Polyethylene, a crystalline plastic, is opaque but a small amount of melt drool out of the nozzle tip will always be transparent.

2.5 Thermal Transitions in Polymers

Although there are no polymers that are completely crystalline, for the sake of the following discussion, let us assume that they do exist. Therefore in the following discussion, a crystalline polymer would mean a polymer that is 100 % crystalline, an amorphous polymer would mean a polymer that is 100 % amorphous and a semi-crystalline polymer would be a polymer that is partially crystalline with the crystallites present in the amorphous regions.

First consider an amorphous polymer submerged in a very cold liquid such as liquid nitrogen. The temperature of liquid nitrogen is anywhere between –210 to –195 °C (–345 to –320 °F). At this temperature, the different molecular energies are almost nonexistent and therefore the molecules are not free to move, resulting in a brittle polymer. A sheet of flexible plastic, when quenched into liquid nitrogen, becomes brittle. If dropped on a concrete floor, it would sound like glass and shatter into bits and pieces. However, if the temperature of the sheet is gradually increased, the thermal energy from the increasing temperature provides energy to the molecules. Depending on the polymer the sheet is made of, at a given temperature, the sheet becomes flexible. This is called the *glass transition temperature* or the T_g of the polymer. At T_g, the molecules have sufficient energy to move and the polymer is flexible. A further increase in the temperature provides more energy, the flexibility increases, and eventually the molecules become soft enough to form a viscous mass suitable for melt processing. An increase in the energy of the molecules results in the increase in the specific volume of the polymer. Specific volume is the volume per gram of the polymer. A plot of specific volume versus temperature, like the one shown in Fig. 2.10, shows a linear increase in the volume until an inflection point is reached, where the slope of the line changes. The inflection point reflects the glass transition temperature of the amorphous polymer. Further increase after the glass transition shows a steady increase in the specific volume.

Next, let us look at a crystalline polymer and perform the same experiment. In crystalline polymers, the intermolecular forces are very high and therefore a considerable amount of energy is required to move the molecules away from each other. As the temperature increases, the molecules gain more and more energy but the intermolecular forces prevent any movement. At the temperature where the molecules move away from each other and become mobile, the molecules are very flexible and the polymer is now a molten mass. The crystal-

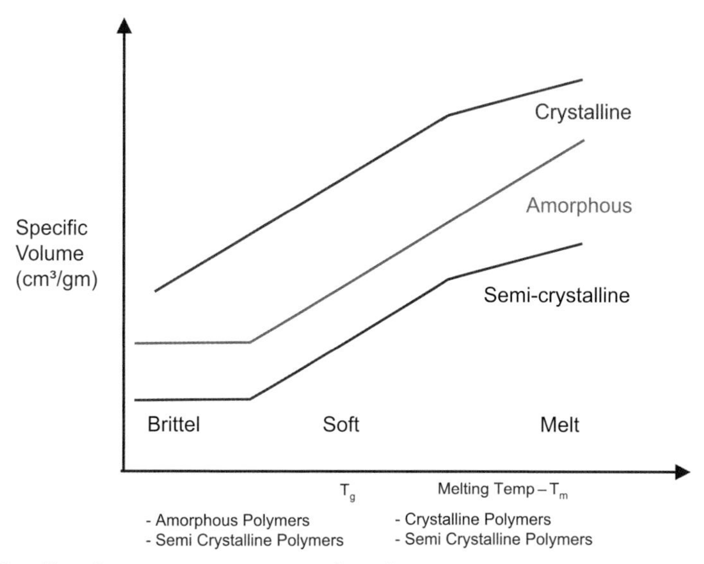

Figure 2.10 Specific volume versus temperature for polymers

lites need a definite amount of energy to melt. When they receive this in the form of thermal energy they melt all at once. The phenomenon is similar to the melting of low molecular molecules, such as water where ice turns into water at 0 °C. For this reason, crystalline materials show a sharp melting point. Because the transition from a solid to a liquid is sudden, the transition is referred to as melting and the transition temperature is called the *melt temperature* or T_m. These polymers do not go through a glass transition temperature. A plot of specific volume versus temperature shows an inflection point at the T_m, as can be seen in Figure 2.10.

Based on the thermal responses of amorphous and crystalline polymers, it is now easy to predict the response of a semi-crystalline polymer. A semi-crystalline polymer can be thought of as crystallites present in an amorphous matrix of molecules. This sample will exhibit the properties of the crystalline and the amorphous polymer and will therefore have a glass transition temperature and a melting point temperature. On the specific volume plot, there will be two inflection points representing each of the transitions, as shown in Fig. 2.11. However, in case of semi-crystalline polymers, the inflection point is not as sharp as one would see in the amorphous or crystalline samples. Rather, there is a range where the inflection starts and ends. This happens because of the different sizes of the crystallites and differences in molecular lengths. Once the molecules of a semi-crystalline material have melted, they now have a lot of thermal energy and any additional energy can start breaking down the molecules causing degradation of the polymer.

This difference between the crystalline and amorphous materials leads to the fact that semi-crystalline materials have a much narrower melt processing window compared to amorphous materials. For example, nylon (a semi-crystalline material) will have a melt processing window between 248 and 265 °C (480 to 510 °F), while an ABS (amorphous material) has a window between 218 and 260 °C (425 to 500 °F). The processing window is 17 °C for nylon, compared to 42 °C for ABS.

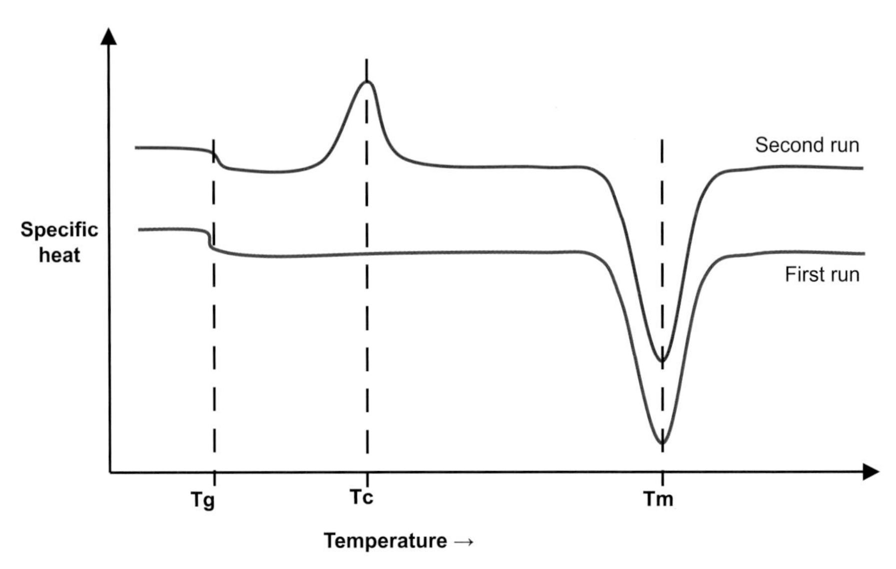

Figure 2.11 Typical representation of a DSC plot for polymers

Now, let us consider the reverse process of phase transitions. When the temperature of a melt of a crystalline polymer is gradually reduced, the molecular energy starts to reduce. The viscosity of the melt starts to increase and the crystallites begin to reappear. The temperature at which the crystallites occur is called the *crystallization temperature* or the T_c. For crystallization to occur, the melt must be subjected to the crystallization temperature for a finite amount of time. If the melt of a crystalline polymer is rapidly quenched below the glass transition temperature, the solidified polymer will not exhibit any crystallites and will be completely amorphous. It is for this reason that mold temperature is of greater importance when molding semi-crystalline polymers as compared to amorphous polymers.

The above discussed temperatures affect the properties of the polymers and therefore decide the final applications of the polymers. For example, for the product designer, the T_g is one of the important factors of consideration. For a product to be flexible at room temperature, its T_g must be below room temperature, such as for elastomers. For a product to be rigid at room temperature, the T_g must be above the room temperature. For the molder, the knowledge about the crystallization and melting temperatures is important because these dictate the processing conditions. The T_c is used to determine the mold temperature range to start and promote the crystallization of the molecules. The T_m is used to determine the melt temperature ranges. A processing data sheet does not provide these as typical values but they are reflected under the processing conditions. The material manufacturer performs the analysis and uses the results as one of the many tests to recommend the processing conditions.

The differential scanning calorimeter (DSC) is an instrument used to determine the thermal transitions in polymers. A typical graph generated by a DSC was shown in Figure 2.11. For crystalline polymers, two scans are performed. During the first scan, the polymer is taken from a low set temperature all the way past the melting temperature, and the transitions are

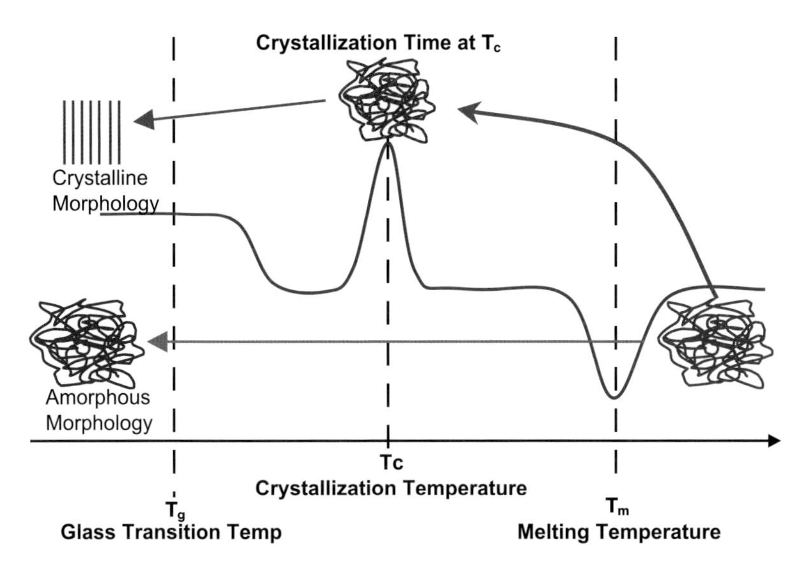

Figure 2.12 Resulting morphologies depending on polymer conditioning

recorded. The sample, which is now in the molten form, is immediately quenched in liquid nitrogen. Most polymers are below the T_g at this temperature. Since the melt is amorphous and is immediately quenched, all the energy is taken away and the frozen polymer is also completely amorphous. The DSC scan is repeated. As the temperature increases, the molecules gain more and more energy. The glass transition is noticed and when the crystallization temperature is reached the crystallites begin to form. Crystallization temperatures are recorded on the second scan. For amorphous polymers, a second scan is not necessary.

The relationship of mold temperature and crystallinity is explained in Section 2.4.4 using the DSC graph, see also Figure 2.12. The figure shows the difference in the resulting morphologies when a polymer sample is quenched below the T_g and when a polymer sample is subjected to the crystallization temperatures for a given time.

2.6 Shrinkage in Polymers

Shrinkage occurs when the melt begins to cool and the molecules start to return to their desired equilibrium states. The distance between the molecules is higher in the melt than when they are cooled. As the melt cools, the molecular distance decreases, reducing the free volume and causing shrinkage. The higher the increase in volume during the melting phase, the higher is the shrinkage. The shrinkage of the plastic during the injection molding process can be easily affected by the various molding parameters. This poses the biggest challenge to mold makers when sizing the core and cavity steel and to the processors in order to maintain the part quality during production. It is therefore important to develop a robust and stable process that will produce parts with consistent shrinkage and that is least affected by natural variations. The shrinkage of injection molded parts also depends on the direction of plastic flow, see Figure 2.13. The plastic is injected through the gate and the molecules get oriented in the direction of flow. Given a chance to relax and with sufficient available energy the molecules will get back to their original non-oriented equilibrium state. But this is not always the case, because molding cycles need to be fast and wall thicknesses are small, inhibiting relaxation. In addition, typically the molecules are under mechanical stress from the injection, pack and hold forces. This causes a variation in shrinkage between the direction of flow and the direction perpendicular to flow. The effect is more pronounced for crystalline materials because they have a high shrink value caused by the formation of the crystallites. Amorphous materials too exhibit a difference in shrinkage depending on flow direction, but the difference is not as pronounced. Materials exhibiting varying shrinkage values in the cross and parallel flow directions are known as anisotropic materials; materials with identical shrinkage values are called isotropic materials. For example, polyesters are anisotropic materials while ABS is an isotropic material. For Valox 357, a polyester manufactured by Sabic Innovative Plastics, the parallel flow shrinkage values range from 1.0 to 1.4% and the cross flow values range from 1.2 to 1.6% . Although there is an overlap, the average values for the cross flow are higher. For Starex AB-0760, an ABS manufactured by Samsung, the shrinkage values in the cross and parallel flows are both 0.30 to 0.60 %.

Exact shrinkage values can never be predicted because shrinkage is a function of various parameters. Some manufacturers publish the shrinkage values for different sample thicknesses because, as stated earlier, in thicker parts heat can be retained for a longer time allowing the molecules to relax and causing more shrinkage. For this reason it is difficult to determine the exact mold cavity dimensions that will yield a molded part with the exact finished dimensions. It is very rare to build an injection mold cavity and cores that are the exact size of the desired finished part or CAD model. Mold makers will typically build the mold 'steel safe' and adjust the steel after the initial sampling of the mold. Another factor that complicates this situation is the fact that final shrinkage is a function of both the cross and parallel flow shrinkages. Rarely will a required dimension be perfectly parallel or perfectly perpendicular to the melt flow. The final shrinkage value is a combination of the two numbers. In this case, leaving the mold cavity and cores steel safe is always a good idea. It is not possible to do so in all cases and the tooling engineer must rely on past experience and make the best decision.

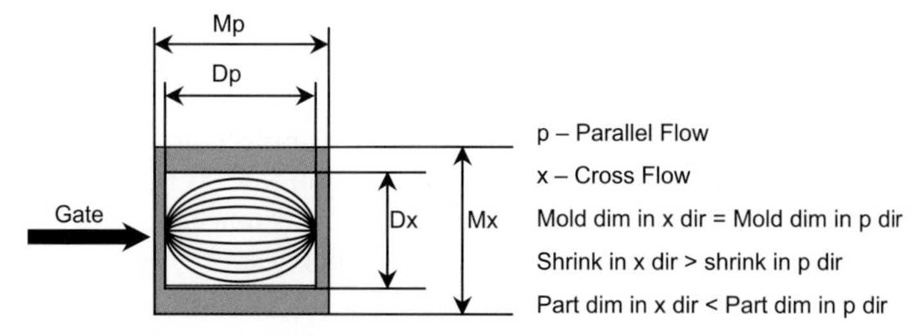

Figure 2.13 Parallel flow and cross flow shrinkage

The term shrinkage in injection molding always compares the part dimensions to the dimensions of the mold. Shrinkage of the melt in a free state is always positive. This means that the volume will always decrease and the part dimension will therefore always become smaller. The term free state is used to describe a non-inhibited state of movement where the molecules are not held by any other forces. However, in some parts, such as long tubular parts, we find that as the part shrinks the length will decrease, forcing the diameter to increase, see Figure 2.14. This can be visualized similar to Poisson's effect where contraction due to external force in one direction leads to the expansion in the perpendicular direction. This negative shrinkage in the diameter of the part is due to the force that is exerted because of the overall reduction in the length of the part which can be significant. Stress can easily build up in such parts and cause premature product failure.

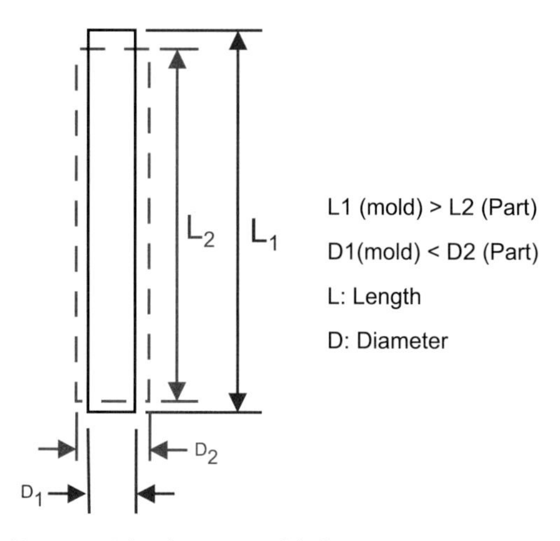

Figure 2.14 Positive and 'negative' shrinkage in molded parts

The phenomenon does not necessarily lead to a negative shrinkage value and can also result in a reduced shrinkage value, depending on the amount of mechanical stress in the part. This poses another challenge for the mold maker in sizing of the cavities. Mechanical stresses can cause a change in dimension in parts with other geometries, such as square boxes with deep pockets. The deep pockets create unsupported walls. Sometimes reduction in the distance between the walls causes an increase in the distance in the other direction. This is represented in Figure 2.15.

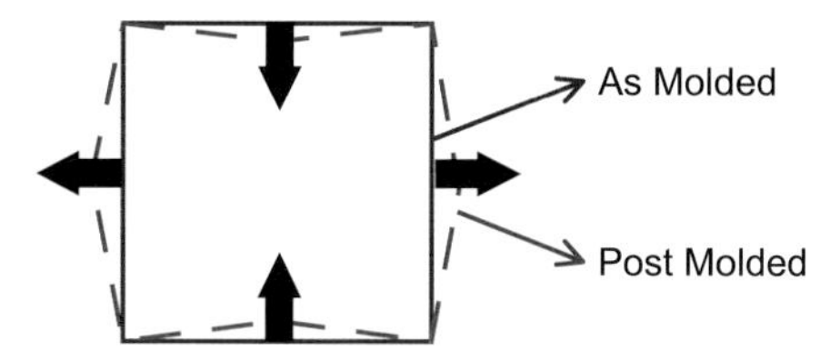

Figure 2.15 Mechanical stresses affecting part dimensions

References

1. Turner, A. and Gurnee, E.F., Organic Polymers, Prentice-Hall, 1967, p 51.
2. Zweifel, H., Maier, R., Schiller, M. , *Plastics Additives Handbook* (2009) Hanser, Munich

Suggested Further Reading

1. Deanin, R.D., Polymer Structure, Properties and Applications (1972) Cahners, Boston, MA
2. Tager, A.A, Physical Chemistry of Polymers (1978) Mir Publishers, Moscow
3. Odian, G., Principles of Polymerization (1991) Wiley Interscience, United States of America
4. Gowariker, V.R., Viswanathan, N.V., Sreedhar, J., Polymer Science (1996) New Age International (P) Limited, Delhi
5. Billmeyer, F.W., Textbook of Polymer Science (1984) Wiley Interscience, NY
6. Brydson, J.A., Plastics Materials (1995) Butterworth-Heinemann Ltd, Oxford, UK

3 Polymer Rheology

Polymer rheology is the science of flow of polymers. The study of polymer flow is essential to understand the melt processing of plastics. In any melt processing technique, the plastic has to be melted and then deformed to conform to the final product specifications. In injection molding, the melt must be injected into a mold and then cooled to get the final part. In extrusion, the melt is shaped through a die and then cooled down to get the required profile. During the injection of the melt into the mold and until it reaches its final form, the melt is subjected to various forces including mechanical and thermodynamic forces. In this chapter, some of these concepts are explained in brief to give the reader an understanding sufficient enough to help apply them to injection molding. The concepts also have been simplified and the general mathematical details have been left out. For those readers, who would like to get an in depth understanding on this topic, a number of books are mentioned in the references at the end of the chapter.

3.1 Viscosity

Viscosity is the resistance to flow. The higher the resistance to flow, the higher is the viscosity. Honey or corn syrup do not flow easily and therefore have high viscosities. On the other hand, water flows very easily and therefore has low viscosity. Gasses have even lower viscosities as compared to water and therefore flow easier. Viscosity depends on a number of factors and is an inherent property of the fluid but can be influenced by external forces. Based on the types of forces the polymer is subjected to and/or on the type of medium the polymer is present, various types of viscosities are defined. Of particular interest in injection molding is the viscosity of the melt as defined by the apparent viscosity. As will be discussed later in the chapter, the viscosity of the melt is dependent on the applied force and therefore is not constant. Hence the adjective 'apparent' is used to describe the viscosity at a particular given shear rate.

Consider a liquid sandwiched between two metal plates as shown in Fig. 3.1. The distance between the plates is H. The area of the plates is A. The bottom plate is held stationary and the top plate is displaced in the X direction with a force F and a velocity $V = V_H$, see Fig 3.2.

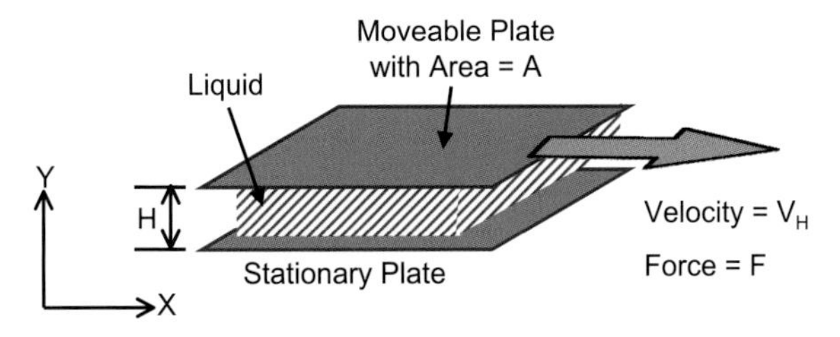

Figure 3.1 Experimental setup to determine the velocity profile

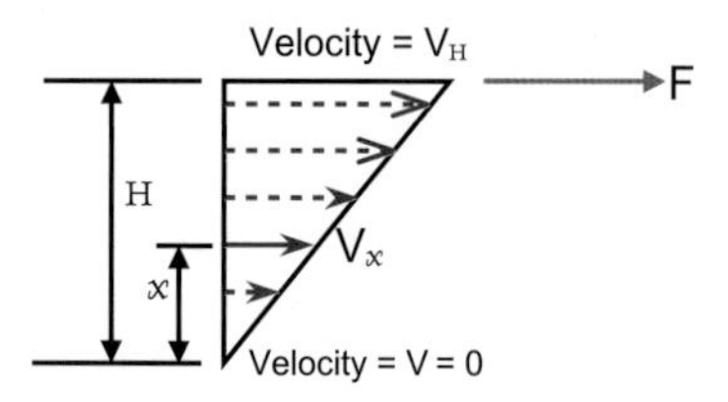

Figure 3.2 Velocity profile for a liquid under shear force

The layer of liquid just below the top plate will also move with the plate at a velocity V_H and the layer just above the bottom plate will be stationary with a velocity equal to $V = V_0 = 0$. At any layer between the two plates, at a distance of x, the velocity is V_x and is proportional to the distance from the bottom plate. The shear rate is the differential velocity between the layers. Since each of the layers is moving at different velocities, the shear rate is also a function of the distance x. The shear rate of the top layer is given by the equation

$$\dot{\gamma}_H = V_H/H \quad (3.1)$$

and at any given layer at a distance x is given by

$$\dot{\gamma}_x = V_x/x \quad (3.2)$$

The units of shear rate are reciprocal seconds or s^{-1}.

Each layer is subjected to shear forces as it is being pulled in the direction of the applied force. Since stress is force divided by area, the shear force τ at the top layer can be defined as

$$\tau = F/A \quad (3.3)$$

Any layer at a distance x from the bottom layer will be subjected to a force F_x, which is between zero and F. The area on which the force is applied stays the same. Therefore the shear stress for this layer will be

$$\tau_x = F_x/A \quad (3.4)$$

F_x and V_x are both proportional to the distance of the layer from the bottom plate. That means the greater the distance, the higher the velocity and the higher is the shear force. The increase in shear stress is directly related to the increase in shear rate. The two are governed by the equation

$$\tau = \eta\dot{\gamma} \quad (3.5)$$

where η is the constant of proportionality called the viscosity of the liquid. The viscosity is the slope of the line when the shear stress is plotted versus shear rate.

3.2 Newtonian and Non-Newtonian Materials

In the above discussion, the shear stress was linearly proportional to the shear rate with viscosity being the constant of proportionality, see Fig 3.3. In some fluids, the relationship between the shear stress and the shear rate is not linear. The viscosity is not a constant and is dependent on the shear rate or the time the fluid is subjected to the shear. Such fluids are not non-Newtonian fluids. Based on the response to shear, non-Newtonian fluids can be classified into two types, dialatent and shear thinning fluids. In dialatent fluids, the viscosity increases with increasing shear rate and in shear thinning fluids, the viscosity decreases with increasing shear rates, see Fig 3.4a. Based on the response to the time of shear at constant shear rate, non-Newtonian fluids are classified into rheopectic and thixotropic fluids (see Fig. 3.4 b). All plastics show shear thinning behavior.

Rheology is the science of the flow of non-Newtonian materials. All polymers are non-Newtonian, specifically shear thinning, as the shear rate increases, the viscosity drops.

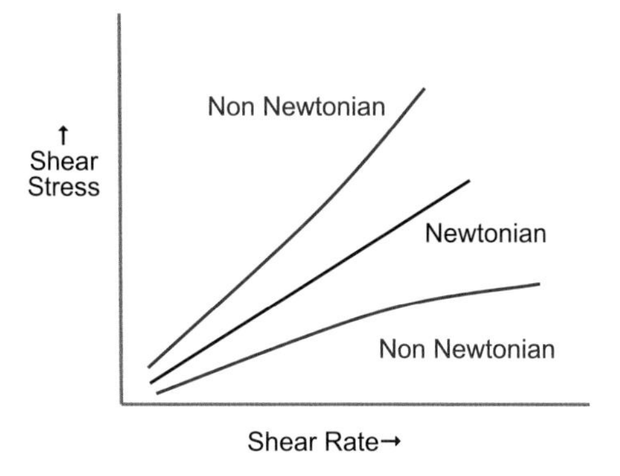

Figure 3.3 Shear stress versus shear rate for Newtonian and non-Newtonian liquids

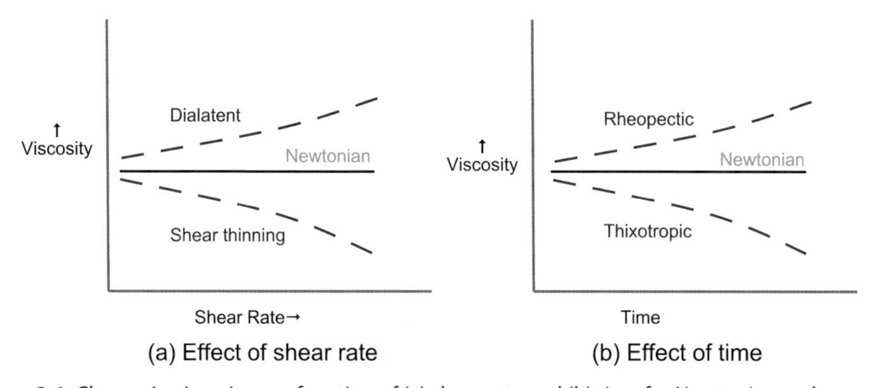

Figure 3.4 Change in viscosity as a function of (a) shear rate and (b) time for Newtonian and non-Newtonian liquids

3.3 Viscosity in Polymer Melts

Polymer melts are non-Newtonian. Research has shown that the velocity profile, which is the line that joins the velocities of each layer, is never a straight line as shown in Fig. 3.2. A typical velocity profile for a polymer melt, called fountain flow, is shown in Fig. 3.5. The parabolic velocity profile is a result of a pressure driven flow through a closed channel. The drag from the channel walls slow the material, resulting in a higher velocity in the center. At the flow front, this condition causes the faster flowing material in the center of the channel to "fountain" to the slower flowing regions. Fountain flow conditions only exist at the flow front and cannot occur behind the flow front. However, the velocity profile conditions exist throughout the flow, at the flow front and behind the flow front., i.e., the center is always flowing faster than the perimeter. Hot runners, where the channel is already full, will not experience fountain flow, except when they are first filled. Fountain flow will exist in the cavities and in cold runners. An incompletely filled runner or a short shot will always exhibit their flow front velocity profiles as shown in Fig. 3.5.

Since the profile is not linear, the shear rate equation in Eq. 3.1 now needs to be rewritten as

$$\dot{\gamma} = (x/h)(dx/dt) \tag{3.6}$$

Where t is time, h is the thickness or diameter of the flow channel and x is the distance along the flow channel. The units do not change and are reciprocal seconds or s^{-1}.

The relationship between viscosity and shear rate is described by various mathematical models. The most popular model, the Power Law Viscosity Model proposed by Ostwald and de Waale, is widely applicable to injection molding. This equation accurately represents the shear thinning region found at the high shear rates in injection molding. The power law viscosity model is given by the equation

$$\eta = m\dot{\gamma}^{n-1} \tag{3.7}$$

where m is a constant called the consistency index and n is the power law index. A representation of a typical graph of viscosity versus shear rate is shown in Fig. 3.6 for a polyester. Note that the shear rates represented here are of the order of those that are experienced in injection molding.

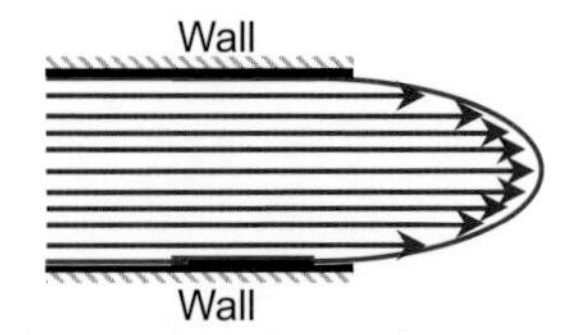

Figure 3.5 Velocity profile of plastic flow

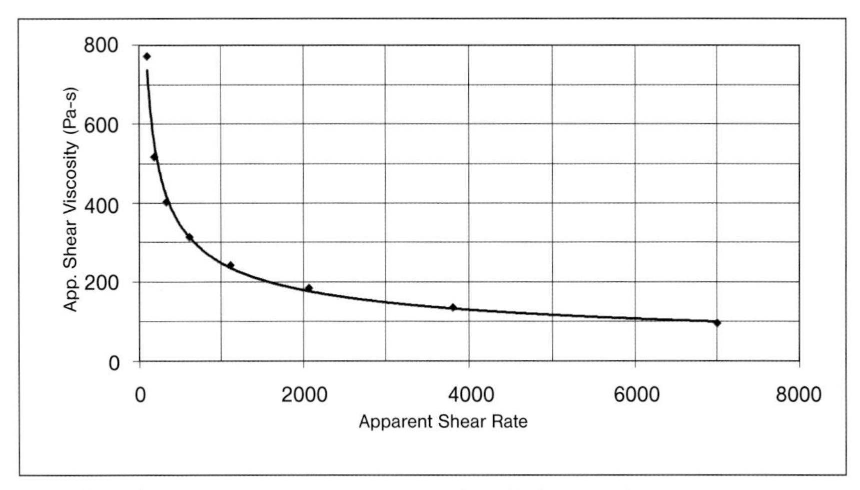

Figure 3.6 Effect of shear rate on viscosity represented on the linear scale

For polymer melts subjected to very low shear rates, the viscosity is essentially unchanged. There is very little or no effect of the shear rate. However, these shear rates are not encountered in injection molding. This is the reason why there are different viscosity models to represent viscosity.

Equation 3.7 can be rewritten as

$$\log \eta = \log m + (n - 1)\log \dot{\gamma} \tag{3.8}$$

This equation now becomes a linear equation and if a graph of log η versus log is plotted, we get a straight line as shown in Fig. 3.7. This is the typical graph that is obtained from a melt rheometer test. The graph in Fig. 3.7 is generated from the same data used for the PBT graph in Fig. 3.6. The slope of the line is $(n - 1)$. Polymer melts are shear thinning and therefore the

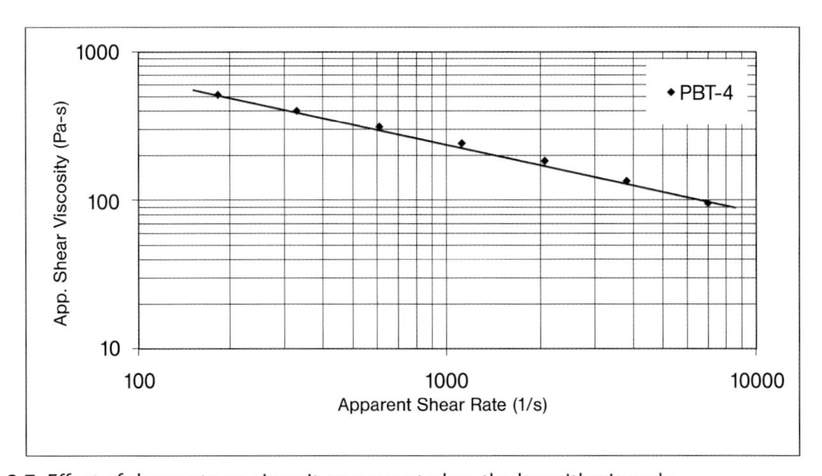

Figure 3.7 Effect of shear rate on viscosity represented on the logarithmic scale

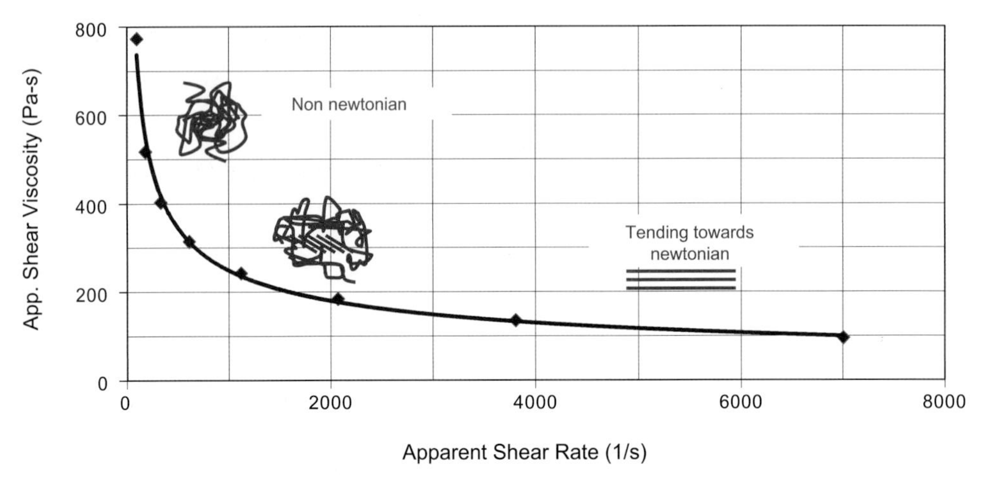

Figure 3.8 Effect of apparent shear rate (injection speed) on the viscosity of the polymer

viscosity drops, resulting in a negative slope of the line. Therefore, the value of n is always less than 1 but greater than zero.

In polymer melts, as the shear rate is applied, the molecules start to align themselves in the direction of flow, moving away from their equilibrium intertwined states. Increasing shear rates stretch and align more and more molecules in the direction of flow. This alignment facilitates the easy movement of the flow layers past each other, thus reducing the resistance to flow or viscosity. At a certain point, all the molecules become aligned in the flow direction and increasing the shear rate further has little or no further effect on the viscosity.

For the sake of practical injection molding, we could consider the region of higher shear rates as a Newtonian region where the viscosities become consistent. Since the viscosities are a result of the fill speeds, the corresponding regions of fill speeds are now treated as the consistent region. Injection speed is synonymous to shear rate and an in-mold viscosity curve appears similar to the one shown in Fig. 3.8. Shear rate can be calculated as the reciprocal of the fill time, where fill time is the time the screw takes to travel from the set shot size to the holding phase transfer position of the screw on the molding machine. Simply put, it is the time for which the screw moves in the injection phase.

3.4 Effect of Temperature on Viscosity

In the solid state the molecules have very little thermal energy and therefore are almost immobile. Depending on the ambient temperature and the glass transition temperature (T_g) of the polymer, the polymer can be either brittle or soft and tough. The thermal regions are explained in Chapter 2. Below the T_g, the plastic is very brittle and is said to be in the glassy state. Above the T_g the plastic is soft and is said to be in the rubbery or viscoelastic state. Above the melting temperature (T_m), for crystalline plastics, the plastic is in the melt form.

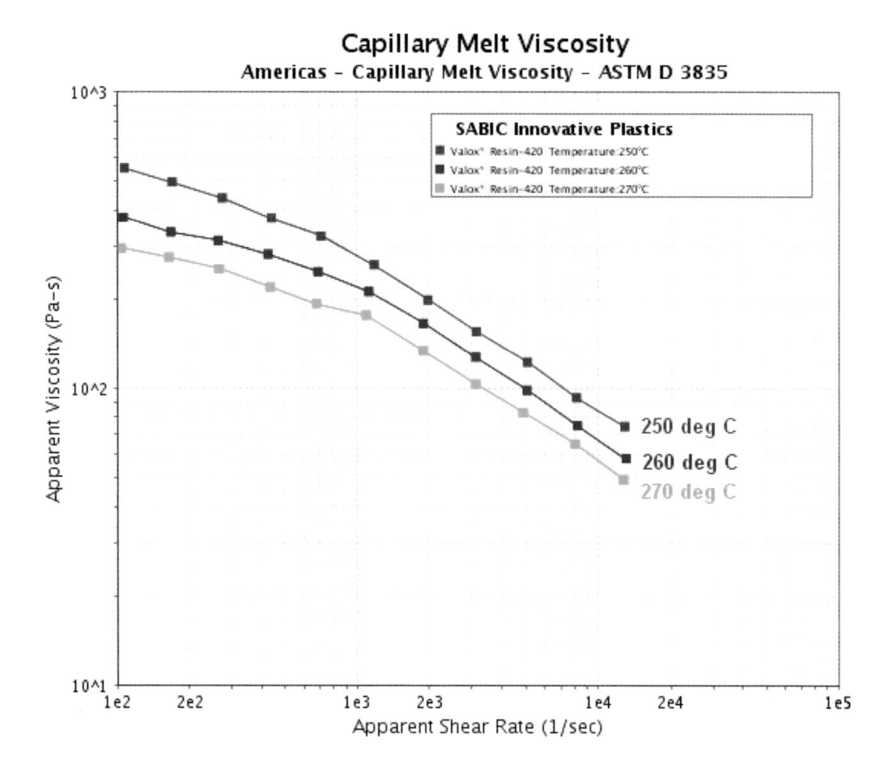

Figure 3.9 Effect of temperature on the viscosity of the melt (Source: Sabic Innovative Plastics)

In general, as the temperature is increased, the thermal energy reduces the weak intermolecular attraction that holds the molecules together, making them more mobile. Amorphous polymers continue to soften and crystalline polymers show a sharp melting point. Increase in temperature increases the mobility of the molecules, thereby reducing the viscosity of the polymer. Temperature and viscosity are inversely related. Figure 3.9 shows the effect of temperature on the viscosity of the melt. It is also clear that the effect of shear rate is higher than the effect of temperature.

In injection molding, it is common practice to increase the melt temperature to make the plastic flow easier. However, Fig. 3.9 demonstrates that increasing the injection speed will have a greater effect on part fill. This advantage is discussed in Chapter 7 and a procedure to generate the in-mold rheology curve is also discussed.

3.5 Velocity and Shear Rate Profiles

The velocity profile shows the relative velocities of the different layers in the polymer melt as it flows through a channel. The length of the arrows in the velocity profile represents the velocity of each of the layers. The velocity at the wall is zero and therefore the shear rates are very low near the wall. The velocity increases towards the center of the channel, following a parabolic profile and it is highest at the center. As discussed earlier, shear rate is the difference in the velocities of the adjacent layers. In Fig. 3.10, the difference in the velocities in the first two layers near the wall is very high compared to two layers near the center of the channel. Therefore, the shear rate is higher near the wall, compared to the center of the chan-

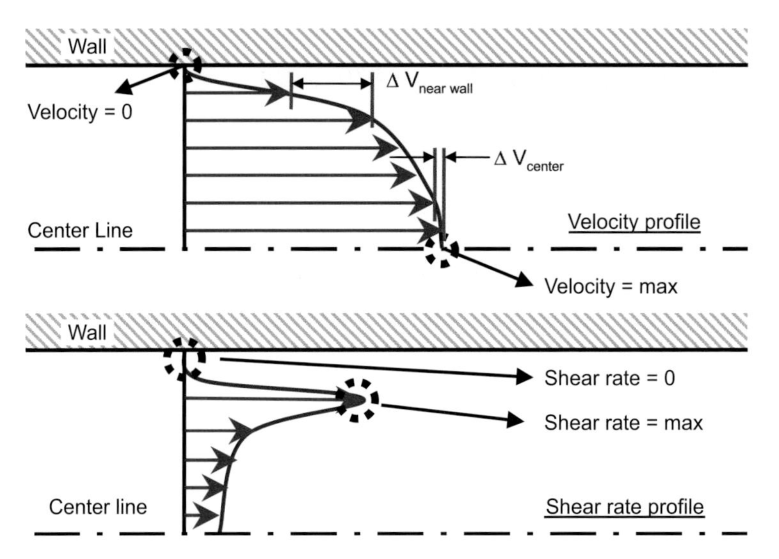

Figure 3.10 Velocity and shear rate profiles in polymer melts

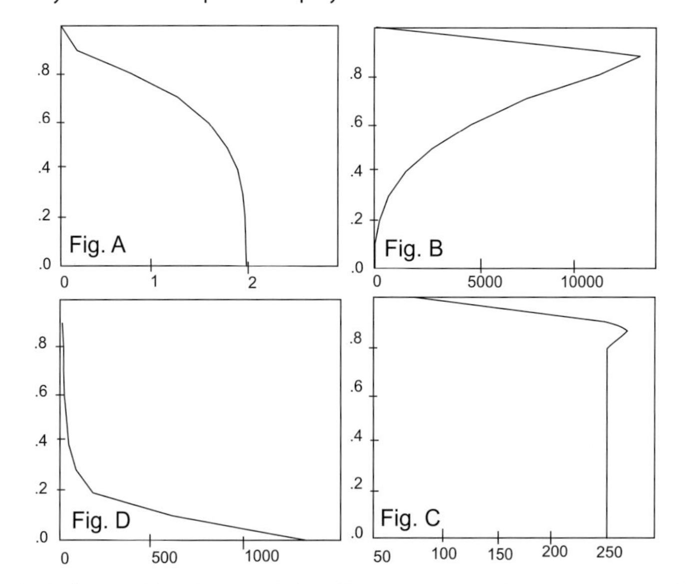

Figure 3.11 Outputs from a finite element, finite difference flow analysis program providing information on the melt conditions through the cross-section of a cold runner. Y-axis is from the center line of the channel to the channel wall: (A) velocity; (B) shear rate; (C) melt temperature; (D) viscosity [1]

nel. The shear rate profile, which is a derivative of the velocity profile, is plotted below the velocity profile in Fig 3.10. These studies are relatively recent [1].

Figure 3.11 [1] shows the outputs from a finite difference flow analysis program. It shows the velocity, shear rate, melt temperature, and viscosity of the cross section of the melt. The y-axis represents the direction perpendicular to flow.

3.6 Application to Injection Molding

The direct effect of the shear rate distribution discussed above is visually evident in injection molding, particularly in the filling pattern of multi-cavity molds. As the plastic begins to flow in the runner, the shear layers are formed as shown in Fig. 3.10. Since the melt flow is always laminar, these layers split and/or flow into the various flow channels in laminas. Each of these laminas has their own characteristic properties, such as shear rates and temperatures. The high shear laminas just below the wall of the flow channel create a low viscosity region changing the velocity of flow in some cavities and causing cavity-to-cavity flow imbalances. Some of the typical and common characteristics of such flows and their effect on the parts are described in the following.

3.6.1 Flow Imbalance in an 8-Cavity Mold

Consider an 8-cavity mold as shown in Fig. 3.12. As the plastic flows through the primary runner, the shear layers are developed.

If we disregard the frozen layer (in a cold runner mold), we can distinguish between two distinct layers. The outside layer is the high-shear layer and the inside layer is the low-shear layer. In the diagram, the high shear layer is the shaded area. Section A-A is the cross section of the primary runner and shows the two layers concentric to each other. Since the flow

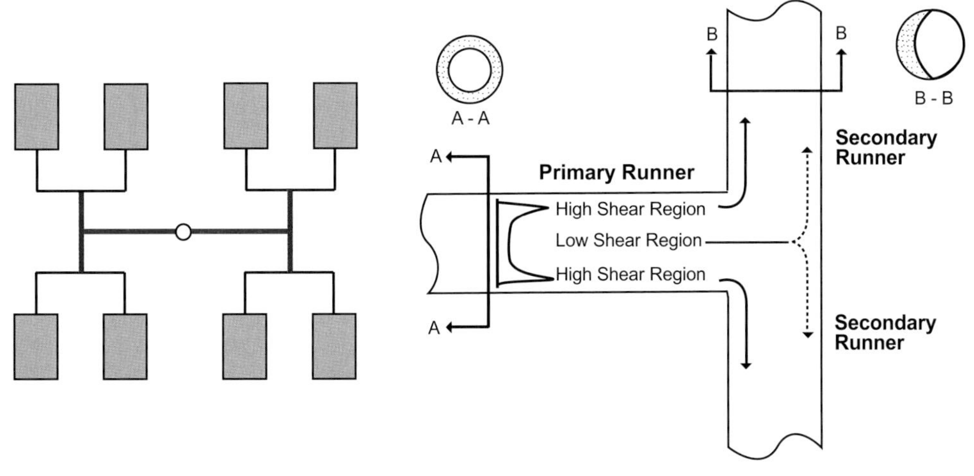

Figure 3.12 Split of different shear rate regions in an 8-cavity mold [2]

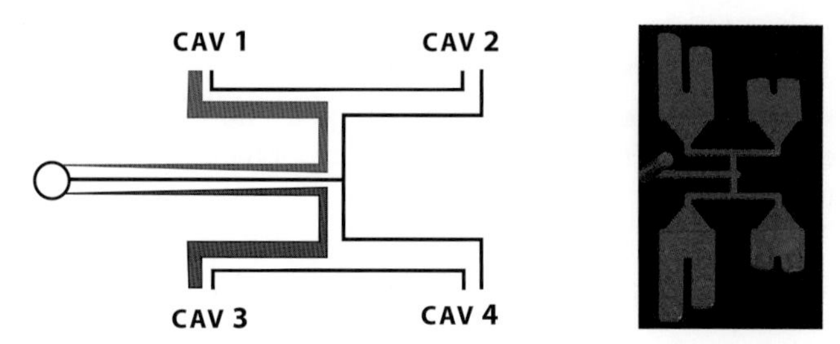

Figure 3.13 Flow imbalance between the inside and outside cavities. The photo on he right is an actual short shot of a polycarbonate material molded at optimal process conditions [3]

through the mold is laminar, the variations in shear, temperature, and viscosity across the runner proceed into the secondary runner. The inner low-shear laminas hit the far wall of the secondary runner and the high-shear laminas on the outer perimeter continue to flow along the near wall of the branching secondary runner. The cross section shown in Section B-B illustrates this distribution of high- and low-sheared material in the branching secondary runner.

The laminar flow continues through the secondary runner, the tertiary runner, and then into the cavities. The result is that inside cavities (those close to the sprue) are filled first, as they are fed by the hotter, high-sheared, lower viscosity material developed earlier in the runner. This is shown in Figure 3.13. A visual proof of the high-shear lamina is shown in Figure 3.14.

Here, a runner used in the molding of a PVC part clearly showed degradation caused by the high shear. The black streak on the inside is the high-shear lamina in the figure. Note that the burning/streaking of the material develops prior to the actual corner. We point this out in particular, because shear induced imbalances are sometime misrepresented as being caused by a sharp corner in a runner. Here, the evidence dispels this theory as the runner is not only radiused (no sharp corners) but the burning begins before the corner.

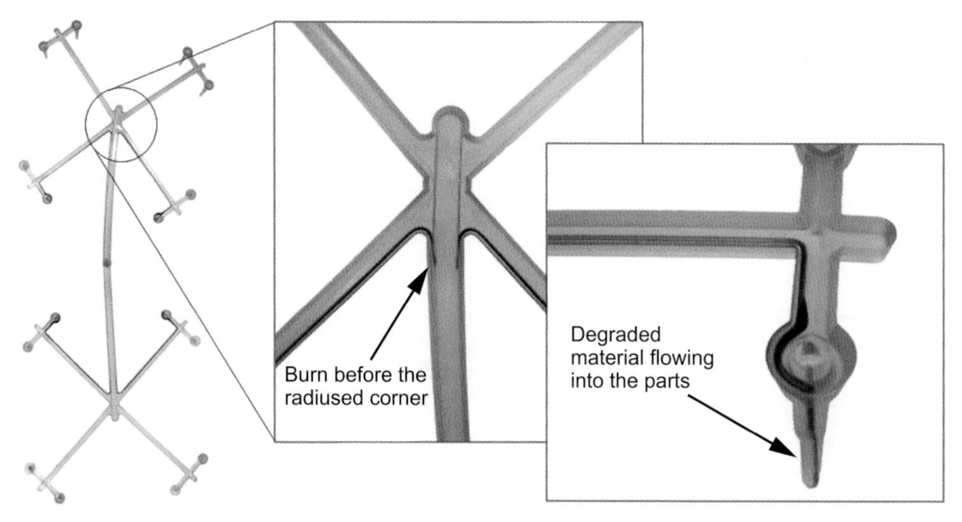

Figure 3.14 Visual evidence of the high-shear lamina [3]

The cavity-to-cavity imbalance described above is called a rheological imbalance. The runner is said to be rheologically imbalanced despite the fact it is geometrically balanced. When the distance from the sprue to the gate is the same for each cavity, the mold is said to be geometrically balanced. This geometrical balance in a runner is still commonly incorrectly referenced as a "naturally balanced" runner.

3.6.2 Racetrack Effect in a Part with Constant Thickness

The part shown in Fig. 3.15 is made with a single-cavity mold, producing a 100 mm 2 mm thick flat part with a constant wall thickness. The shear effect on viscosity can be seen in the flow pattern developed in this cavity. The high-sheared low-viscosity material developed in the perimeter of the runner splits and concentrates along the perimeter of the flat disc, causing the race tracking effect. Note that the photo in the right shows that the effect is significant enough to create a gas trap, opposite the gate, in this flat part.

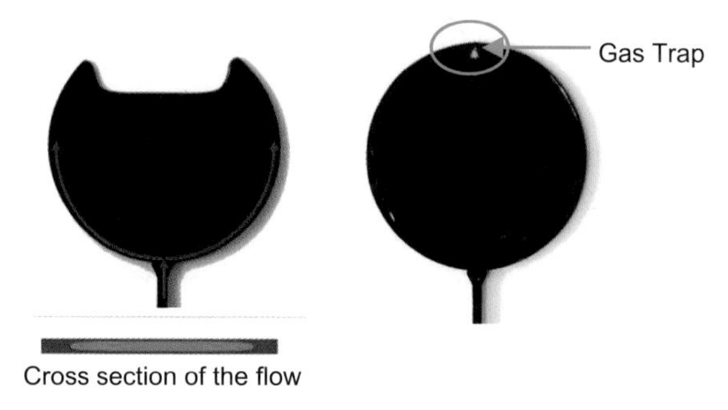

Figure 3.15 Racetrack effect causing a gas trap in a part with uniform thickness [3]

3.6.3 Stress Build-Up in Molded Parts

The part shown in Fig. 3.16 is made from a transparent material in a two-cavity mold. The parts are packed out and examined with a polarizing lens after molding. The build-up of stress can be seen on the inside of the parts. This is the area where the hotter laminas flow, causing differential cooling and therefore stress.

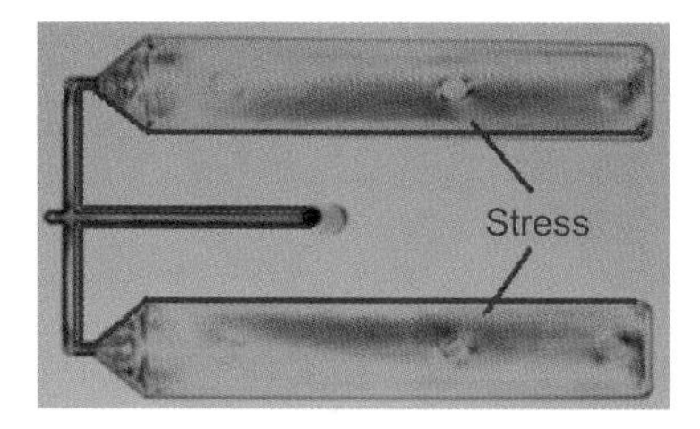

Figure 3.16 Stress build-up observed under a polarizing lens [3]

3.6.4 Warpage Difference Between Cavities

Figure 3.17 shows parts molded in a 4-cavity mold. Because the location of the hot laminas in the cavities is different, two of the cavities are warped while the other cavities are perfectly flat.

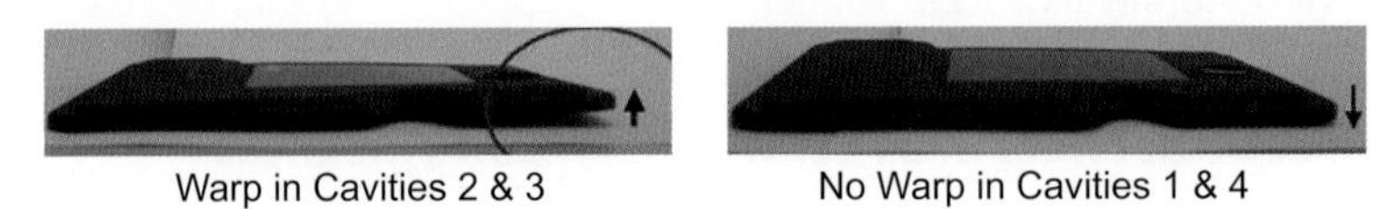

Figure 3.17 Warpage differences between cavities from the same mold caused by a melt imbalance [3]

3.7 Solving Flow Imbalances Using Melt Rotation Techniques

The solution for balancing the flow and creating rheologically balanced molds was developed and patented by John Beaumont of Beaumont Technologies in Erie, Pennsylvania (please note that the use of this technology requires a licensing agreement from Beaumont Technologies Inc.) Beaumont's varied methods of melt management, commonly known as Melt-Flipper® technology, can be used in applications that include the rheological balance of mold and part filling, control of intra-cavity filling, warpage, part property, and cosmetic manipulation. One of the more common applications is shown for an 8-cavity mold in Fig. 3.18 a. This is a conventional H-shaped runner with eight cavities. The cross sections of the flows are also shown. In the primary runner, the high shear and the low shear areas are concentric to each other. As the flow splits at the secondary runner, the high-sheared material stays on

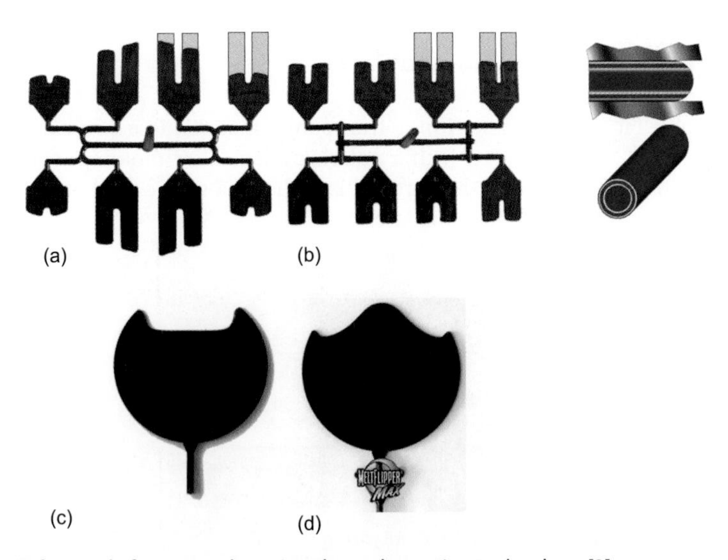

Figure 3.18 Before and after examples using the melt rotation technology [3]

the inside. At the split at the tertiary runner, the high-shear material ends up in the inside cavities, causing these cavities to fill before the outside cavities. Inside and outside cavities exhibit different melt conditions. The result is that the parts formed in these two cavity groups will be different in size, weight, and properties.

Beaumont's patented melt rotations technologies use a variety of methods to manage the position of the high- and low-sheared laminates to achieve the desired balancing effects. In this 8-cavity example, the melt would be "flipped" or rotated, 90 degrees prior to entering the tertiary runner to the position shown in Figure 3.18b. This rotation is commonly created at the intersection of the primary and the secondary runner. When this melt exits the secondary runner, the high-shear area is now on the top rather than on the inside, as is shown in Fig 3.18a without the flip. When this reoriented material enters the split of the tertiary runner, the high- and low-sheared melt splits up evenly into the two branches and each of the cavities receives melt with equal amounts of high-shear and low-shear material. This creates a fill and rheological balance between all cavities in the mold.

In the above case, the melt was rotated at one location, the intersection of the primary and the secondary runner. In case of a 16-cavity mold, the concept can be extended and the melt will need to be rotated at two locations for balancing all sixteen cavities. In addition, by applying similar melt rotation techniques, some of the other problems in the parts can be solved. An excellent treatise on this subject with detailed explanation is provided in [1]. Some of the 'before and after' examples are shown in Fig. 3.18. Figure 3.18a shows the filling pattern resulting in a conventional geometrically balanced runner. Filling is not only unbalanced from cavity to cavity, but each side of the Flow #1 cavities (inside four cavities) is different. Figure 3.18b shows the filling pattern after melt rotation was applied. Note that a balanced filling results not only between cavities but also within cavities. Not only will all eight parts be almost similar, but the use of pressure transducers (or thermocouples) for controlling and monitoring the process can be significantly improved. Figure 3.18c shows the filling pattern resulting in a simple flat disk, where the high-sheared material from the runner is causing a race track effect around the perimeter. Figure 3.18d is the same disk, except the melt has been conditioned using Beaumont's multi-axis rotation technology (MAX™ technology). Here, the majority of the high-sheared laminates have been repositioned to flow across the center region of the cavity.

Figure 3.19 shows the temperature distribution before and after the use of melt rotation technology. The photos of these parts were taken with an infrared camera immediately after molding. The left hand side part is the inside cavity fed by the high-sheared lamina. In the conventional

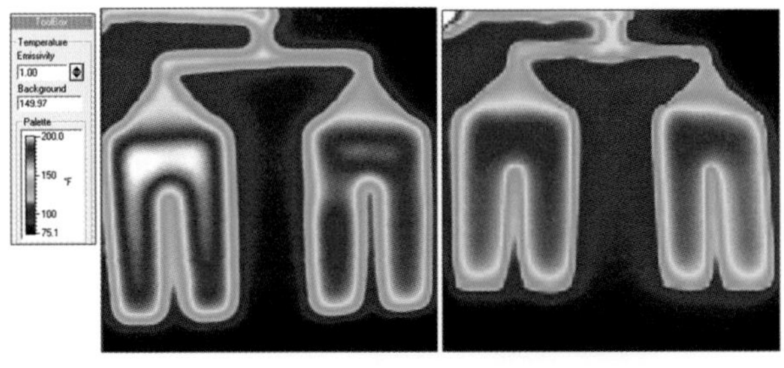

Figure 3.19 Temperature distribution before and after using melt rotation technologies [3]

runner there is clear evidence that the inside cavity is fed with hotter material (white). Note that this evidence of thermal variation exists even after the parts were partially cooled in the mold.

3.8 Characterization of Polymer Viscosity

Polymers are used in a variety of applications ranging from use in a solution (as in the case of paints) to being processed as melts. In these range of applications, the shear rates range from very low to very high. There is a need to employ various techniques to characterize the viscosity for a couple of reasons. First, there is no one viscosity model that fits the wide range of shear rates. A low shear model will not fit the viscosity-shear rate relationship at a high shear rate and vice versa. Second, the presence of another medium, such as a solvent or a plasticizer, alters the models and the properties of these mediums have characteristic effects on the viscosity. Based on these different characterization methods, various viscosities are also defined. *Solution rheology* is the study of polymers in solvents and *melt rheology* is the study of rheology of melts. Of particular interest related to injection molding is melt rheology at high shear rates. This is done with the help of the *capillary rheometer*. The shear rates that are observed in the machine nozzle, sprue, runner, gates, and the part can all be duplicated in this rheometer. Typical shear rates can range from 100 s^{-1} to almost 100,000 s^{-1}, with the shear rates in the gates potentially going into the millions per second. Figure 3.20 shows a schematic of a basic rheometer. It consists of a barrel and plunger arrangement. On the bottom end of the barrel is a die. The barrel is heated to a desired temperature based on the material being tested and is then filled with the plastic to be tested. Once molten, the plunger travels at various shear rates and the plastic is extruded through the die. The geometry of the die is

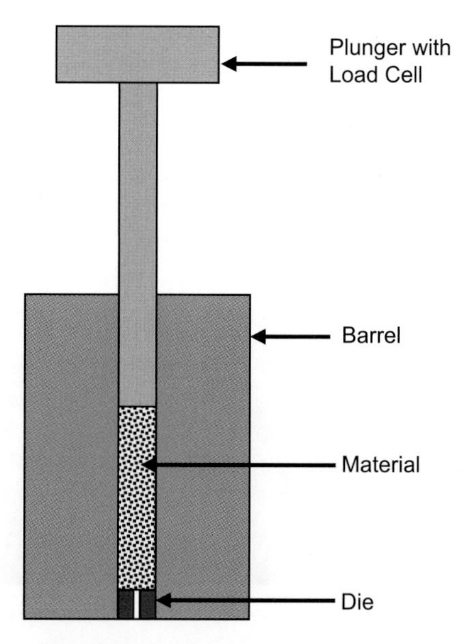

Figure 3.20 Schematic of a capillary rheometer

very important and is used in the calculations of the shear rates. The plastic is extruded at various shear rates and the viscosity is measured. Typically, the viscosity is measured at three different temperatures. These data are is sometimes made available by the material suppliers for mold design and flow simulation purposes.

Another common test is called the *melt flow test*. This test is carried out using a melt flow indexer that has a basic arrangement similar to the capillary rheometer. The difference is that this is a low shear test, with shear rates not commonly seen in injection molding. The advantage of simplicity and a general correlation with the high shear viscosity makes this popular. The output number is called the melt flow index (MFI) or the melt flow rate (MFR). The melt flow indexer is similar to the capillary rheometer, except that instead of the load cell, a known weight is placed on the plunger. The arrangement consists of a plunger, barrel, and a die. The barrel is set to a desired temperature, then loaded with the plastic. Once a predetermined time has lapsed, the weight is placed on top of the plunger. The weight forces the plastic to be extruded. The weight of the plastic extruded in 10 minutes (in g) is called the MFI or simply the melt index of the plastic. For example, if 25 g of plastic was extruded from the die, the MFI is 25. In the industry, it is common to say that the material is a 25 melt material. The units for MFI are always grams regardless of the measurement system. The MFI number is commonly found on material certifications as an incoming quality control parameter. A number of companies perform their own test to confirm these numbers and to keep a log of the incoming lots of material. This is an easy and inexpensive way of quality control, though care must be taken with filled compounds. In this case, the fillers do not melt and can cause an obstruction to flow and bias the results making them inconsistent.

The procedures for the capillary rheometer and the melt flow indexer tests are both defined by ASTM or other test organizations. They also define the various temperatures, weights, times, and other test conditions that may vary based on the type of plastic being tested.

References

1. Beaumont, J., Runner and Gating Design Handbook (2007) Hanser, Munich
2. Beaumont, J., et al., Solving Mold Filling Imbalances in Multicavity Injection Molds, *Journal of Injection Molding Technology*, June 1998, Vol 2, No 2, p. 47
3. Beaumont Inc., Technical Presentation, (2009)

Suggested Further Reading

1. Beaumont, J., Nagel, R., Sherman, R., Successful Injection Molding (2002), Hanser Publications, Munich
2. Aklonis, J.J., Introduction to Polymer Viscoelasticity (1983) Wiley Interscience, NY
3 Billmeyer, F.W., Textbook of Polymer Science (1984) Wiley Interscience, NY
4. Cogswell, F., Polymer Melt Rheology (1981) John Wiley, NY
5. Dealy, J., Wissbun, K., Melt Rheology and its Role in Plastic Processing Theory and Applications (1990) Van Nostrand Reinhold, NY

4 Plastic Drying

Most plastics tend to absorb moisture when exposed to humidity. This is true with plastic in any form, whether in pellet form before processing or in a finished product, such as an injection molded product. Such plastics are called hygroscopic or hydrophilic plastics. Plastics that do not absorb moisture are called hydrophobic plastics. Nylons are common examples of hygroscopic plastics. Nylon parts will absorb moisture and alter the dimensions of the molded part, depending on the humidity. As a nylon part absorbs moisture, it can swell in physical size, causing dimensions to change beyond the required specification limits. Although moisture absorption is inevitable in the molded part, excess moisture should be removed from the plastic resin to an acceptable level before molding in order to produce an acceptable part. Every plastic has an acceptable maximum moisture level above which melt processing problems can occur. Moisture levels must be below this recommended value before processing. A list of maximum moisture levels for various materials is provided in Table 4.1. The numbers mentioned are for non-filled plastics. Most fillers are non-hygroscopic and therefore do not absorb any moisture. Consider as an example non-filled nylon. Nylons typically need to have a moisture level of less than 0.20 % before they are processed. If a particular nylon resin is 50 % glass filled, then the amount of nylon is 50 %. Therefore, the amount of allowable moisture will also be 50 % of 0.20 or 0.10 %. The amount of filler must always be taken into consideration when conducting a moisture test. Most material manufacturers overlook this when providing the datasheets.

Table 4.1 Unfilled Materials and the Recommended Maximum Moisture (Courtsey: IDES.com)

Long name	Short name	Suggested max. moisture (%)
Acetal (POM) copolymer	Acetal copolymer	0.15–0.20
Acetal (POM) Homopolymer	Acetal Homopolymer	0.2
Acrylic, polymethyl methacrylate	PMMA	0.097–0.10
Acrylonitrile butadiene styrene	ABS	0.010–0.15
Polyamide 6	Nylon 6	0.095–0.20
Polyamide 66	Nylon 66	0.15–0.20
Polyamide 66/6 copolymer	Nylon 66/6	0.099–0.20
Polyphthalamide	PPA	0.045–0.15
Polycarbonate	PC	0.019–0.020
Polybutylene terephthalate	PBT	0.020–0.043
Polyethylene terephthalate	PET	0.0030–0.20

Table 4.1 (*continued*) Unfilled Materials and the Recommended Maximum Moisture

Long name	Short name	Suggested max. moisture (%)
Polyether imide	PEI	0.020 – 0.021
Polyethylene, high density	HDPE	NA
Polyethylene, low density	LDPE	NA
Polyethylene, linear low density	LLDPE	NA
Polyphenylene sulfide	PPS	0.015–0.20
Polypropylene homopolymer	PP homopolymer	0.050–0.20
Polystyrene, general purpose	PS (GPPS)	0.02
Polystyrene, high impact	HIPS	0.1
Polyvinyl chloride	PVC	NA
Styrene acrylonitrile	SAN	0.020–0.20

In materials such as nylon, moisture plays an extremely important role during processing by acting as a viscosity regulator for the melt. Therefore, there is also a minimum required level of moisture for such materials. This topic will be discussed further later on in this chapter. Drying plastic resins before processing is a critical step. Plastic resins that are hydroscopic must be subjected to elevated drying temperatures for a specified time, in order to effectively remove excess moisture. However, excessive drying of the plastic resin, beyond the manufacturer's recommended times and temperatures, can also create problems. Until recently, this has been an overlooked condition although it can have a very negative potential impact on the mechanical properties and appearance of the finished molded part. Exposure of plastic resins to drying temperatures above the manufacturer's recommended levels for excessive periods of time is typically referred to as "over-drying".

If the process of drying is not done correctly, it can result in losses in production in the form of scrap parts and also lost production time that cannot be recovered. Drying is accomplished with the help of dryers in the molding facility. Some plastics come pre-packaged in vacuum sealed bags and do not need to be dried as long as they are processed immediately after being removed from their packaging. Any opened and unused portion must be dried before it is processed, if later stored in an open environment.

4.1 Problems in Melt Processing Related to the Presence of Moisture

There are several problems that can result due to the presence of moisture during the melt processing of the plastic.

4.1.1 Degradation of Plastic

In the presence of moisture, the plastic can degrade in the injection molding barrel. A chemical reaction called hydrolytic degradation can take place at molding temperatures and attack the long-chain molecules. The water molecule acts as a catalyst and initiates the degradation. The degradation can itself produce more water and increase the rate of the degradation reaction. This breakdown of the molecules causes the loss of properties in the final product. There are two types of hydrolytic degradations that can occur: If the chain ends are broken, the loss in molecular weight is not significant and the effect on the end product is negligible. This type of degradation is called end-chain degradation. In random degradation, the polymer molecules are broken at random bonds along the molecule's length and this can cause a significant drop in molecular weight, leading to a decrease in properties. Typically, parts molded from degraded materials tend to become brittle and show a reduction in their mechanical properties. There can also be a loss in appearance ranging from a shiny finish to a dull finish. Other surface defects, such as splay, are also common. Hydrolytic degradation is a common problem with condensation polymers, such as polyesters, nylons, polycarbonates, and polyurethanes.

4.1.2 Presence of Surface Defects

A number of surface defects can be attributed to the presence of moisture during processing.

Splay

Any moisture that gets carried by the melt during injection of the melt into the mold stream gets released and then tends to form a film between the melt and the mold surface. The presence of this film prevents the melt from coming in contact with the mold surface and picking up the texture of the cavity surfaces. The film creates a smooth surface and once cooled down, the plastic shows shiny streaks in this area, see Fig. 4.1. This phenomenon is called splay or sliver streaks.

Figure 4.1 Splay on the surface of a part

Bubbles

If the moisture stays inside the melt and does not come out to the surface, internal defects, such as voids or bubbles, can form in the part. In some cases, if the bubble is too close to the surface, the moisture inside is still hot and pressurized when the part is ejected from the

Figure 4.2 Internal voids and external defects on a part due to excessive moisture

mold. This can show up as an external defect such as a bump or bubble on the part. These defects can either be microscopic or in some cases show up as large deformations common in large and thick parts. Figure 4.2 shows internal voids and external defects due to excessive moisture.

Burn Marks

High injection speeds cause the plastic to be subjected to high shear rates. At these shear rates and in the presence of moisture some plastics can undergo degradation and burning of the molecules. This can show up as dark streaks or discoloration in the part. Sometimes, the burning can be seen at the end of fill. Vents are designed to help remove the displaced air in the cavity and runner system as the plastic fills the mold during the mold filling process. If moisture is present in the melt, the plastic can degrade, which causes an excess amount of gas that needs to be evacuated from the mold cavity. Once the moisture vapor gets to the front of the melt stream, it does not mix with the melt and always stays in front of the melt stream. If sufficient venting is not provided to allow the moisture vapor or gas to escape, it will get trapped and compressed under the high injection, pack, and hold pressures, resulting in a diesel effect that causes burning of the plastic. This burning can be seen as a black or a white mark at the end of fill location, depending on the type of plastic.

Plugging up of the Vents

The excessive moisture or the volatiles from the degraded polymer and its additives can overload the venting system of the mold. This degradation generates by-products that are low-molecular weight compounds. These compounds begin to collect in the vents, plugging them up. As production continues, the part quality starts deteriorating because of reduced venting capacity. Burn marks on the parts near the end of fill are common and frequent when this occurs. Cleaning of the vents is required to eliminate the burn marks. Figure 4.3 shows residue built up on the mold surface.

Dimensional Variation

Plugging of the vents can cause dimensional changes in the part. Part dimensions are directly related to plastic pressure in the cavity. Therefore, to obtain consistent quality, the cavity

Figure 4.3 Residue built up on the secondary vents

pressure must be consistent during every injection molding cycle. However, if the vents are clogged, the air and gasses cannot escape and the internal pressure of the cavity increases, producing a part of varying dimensions. Parts with tight tolerances can easily drift out of specification limits.

Loss in Properties at the Weld Line

When resins are not properly dried, moisture will be present at the flow front; when two flow fronts meet, the moisture interferes and does not allow weld strength to develop to its full potential. This causes reduced weld line strength. Sometimes the weld line is also more noticeable in such cases. This loss in weld line strength can also be a result of insufficient venting due to plugged vents as described earlier.

Nozzle Drool

This is a common phenomenon in crystalline plastics such as nylons. Under normal circumstances, plastic does not leak from the nozzle tip after the screw recovers and fills the barrel with resin. There are two main factors that prevent nozzle drool: First, at the end of the screw recovery process the screw is sucked back without rotation for a short distance. This reduces the internal plastic pressure and gets the molten plastic away from the nozzle tip. Second, the tip is controlled at a temperature that keeps the viscosity of the plastic high enough to prevent it from flowing out of the nozzle but low enough to allow plastic flow through the nozzle at the start of the next injection cycle. This control is extremely important and the presence of any moisture in the plastic can reduce the viscosity to an extent that the plastic will continually drool out of the nozzle tip. No amount of reduction of nozzle temperature or increase in suck-back can help because the plastic flow rate is high, preventing sufficient time for the plastic to cool down through the short nozzle tip. Drying of these resins to control the drool is therefore important. Since crystalline plastics have a narrow processing range, they tend to have lower viscosities compared to amorphous plastics. Drying of such resins to control the drool becomes important. When the material is first loaded into the dryer and is stagnant in the dryer for the required amount of time, the material near the outlet of the dryer is usually not exposed to the dry air. This is due to the design of the hopper where the cone that supplies the dry air sits above a certain level. Only plastic that is above this cone is subjected to the dry air and gets dried. It is advisable to drain out the plastic until dry plastic is encountered and then start the molding. In some cases, when the moisture content is very high, it is not uncommon to see sputtering of the melt out of the nozzle. Once the molding process is started, materiel is continuously flowing from the top to the bottom of the hopper and dry material is constantly supplied to the molding machine.

Inconsistency in Shot Control

The presence of moisture can mask the true volume of the plastic during the shot build up and therefore cause a reduction in the amount of the required plastic. Although the screw will always reach the shot size or build the required volume for injection, it may not always be equal to the volume of the shot that is essential to make a good quality part. Every shot will therefore be of a different weight, causing inconsistency from shot to shot.

Examples of hygroscopic materials that do not degrade in the presence of moisture are ABS, SAN, and acrylics. Nylons and polyesters are examples of materials that will degrade in the presence of moisture. Examples of non-hygroscopic materials are polyethylenes, polypropylenes, and polystyrenes. These materials do not absorb moisture; however, in humid environments moisture can settle on the surface of the un-molded pellet, resulting in cosmetic defects, such as splay if the material is not first dried to remove the surface moisture.

4.2 Hygroscopic Polymers

Polymers that absorb moisture are called hygroscopic. Whether a polymer absorbs moisture depends on its chemical structure, see Fig. 4.4. The water molecule shown here is made up of two atoms of hydrogen and one atom of oxygen. Because the two hydrogen atoms and the oxygen atom tend to share electrons, the hydrogen atoms are usually on one side of

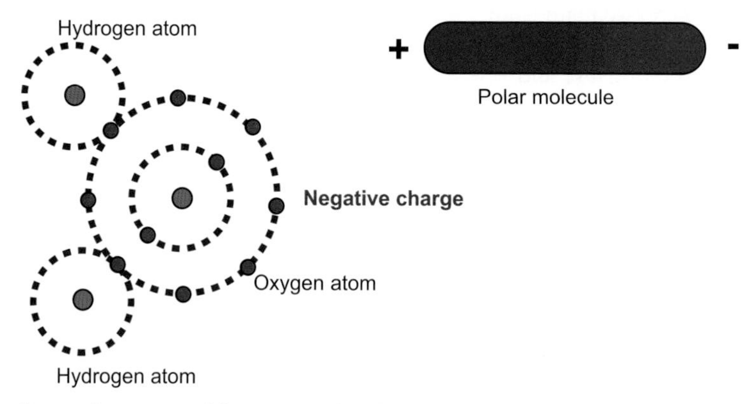

Figure 4.4 Chemical structure of the water molecule

the molecule, which results in the molecule being positively charged on one side and negatively charged on the other. A molecule with this kind of charge distribution is called a polar molecule.

Similar to water, certain groups on the main chain of a polymer can also form polar groups. For example, the carbonyl group (–C=O–) found in polyamides and polyesters is a polar group (see Fig. 4.5). These polar groups on the polymer chains attract the polar water molecules like magnets to form weak secondary bonds resulting in the hygroscopic nature of the polymer (Fig 4.6). These secondary bonds are also called hydrogen bonds.

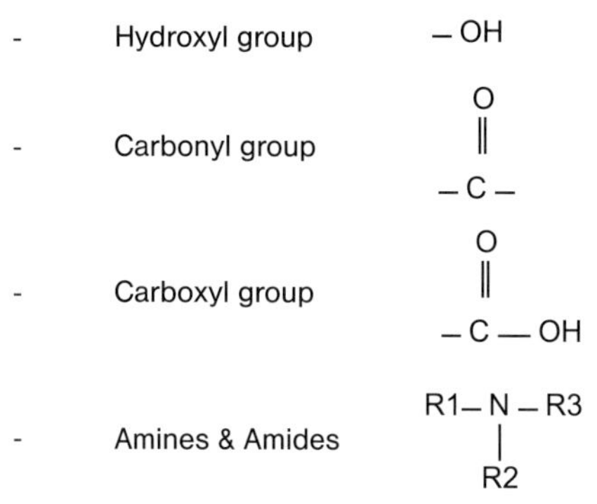

Figure 4.5 Examples of polar groups

Some polymers, for example polyethylenes, are polymerized from non-polar monomers such as ethanes. In an ethane molecule, the charges are balanced around the carbon atoms and hence the polymer is non-polar. The polar water molecule is therefore not attracted to the non-polar polyethylene molecule. Polyethylene is therefore non-hygroscopic (see Fig. 4.7).

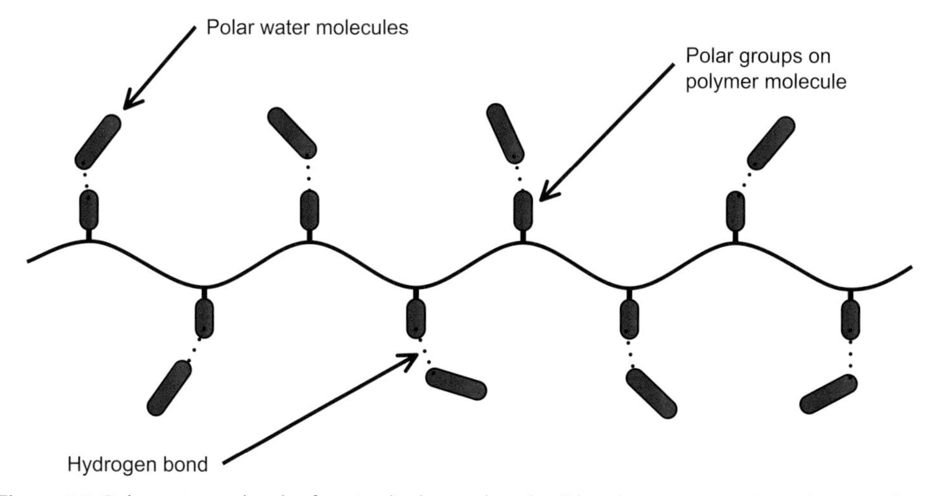

Figure 4.6 Polar water molecules forming hydrogen bonds with polar groups on the polymer molecule making the polymer hygroscopic

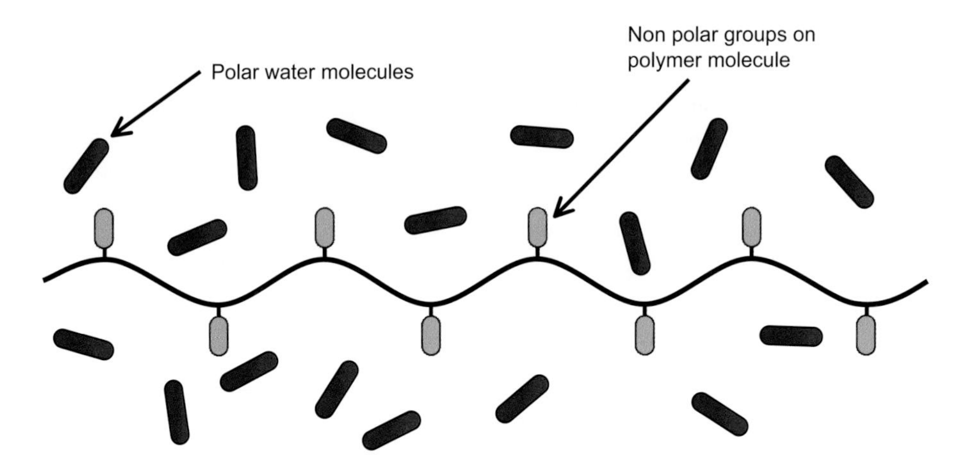

Figure 4.7 Absence of hydrogen bonding between water molecules and polymers with non-polar groups making the polymer non-hygroscopic

4.3 Drying of Plastics

There are some important considerations that need to be taken into account for effective plastic drying before the melt processing stage.

4.3.1 Drying Temperatures and Times

The nature of each polar group for every plastic is different and the strength of the bond between the water molecule and the plastic varies. For example, in polyurethanes the bond strength is very high compared to the bond strength in ABS. For this reason, the drying times and temperatures for individual plastics are different. Depending on the specific grade, polyurethane is typically dried 4 to 6 hours at 132 °C (270 °F), whereas an ABS needs to be dried for 2 to 4 hours at around 75 °C (165 °F). The range in drying times is due to the fact that initial moisture content can vary depending on the humidity the polymer material is exposed to. If moisture content is higher, the drying time required to remove the moisture will be longer, while the rate per unit time of moisture removal stays the same. Only a given amount of moisture can leave the plastic in a given amount of time. The size and the shape of the pellet can also affect the drying times. The drying temperatures and times for some common plastics are listed in Table 4.2. Note that some of these materials may not be injection molding materials.

With some plastics, various combinations of drying times and temperatures can be used to dry the material. For example, polyphthalamide (PPA) can be dried at 79 °C (175 °F) for 8 hours or at 212 °C (250 °F) for 2 hours. The choice of recommended combination depends on the optimal residence time of the plastic in the dryer. Sometimes the dryers can be large compared to the material used per hour. Care must be taken because certain resins can be very sensitive to higher drying temperatures, especially if subjected to longer times in the dryer than suggested by the manufacturer. Prolonged drying times can harm the plastic. In

Table 4.2 Drying Temperatures and Times for Common plastics (Courtsey: IDES.com)

Name	Drying temp.	Drying time
ABS	175 – 190	2.0 – 4.0
ACETAL HOMOPOLYMER	175 – 195	2.0 – 4.0
ACRYLIC (PMMA)	180	3.0 – 6.0
ASA	180 – 190	2.0 – 4.0
NYLON 6	160 – 180	2.0 – 4.0
NYLON 66	175	2.0 – 4.0
PBT	250 – 280	3.0 – 4.0
PC	250	3.0 – 4.0
PEEK	176	3
PEI	300	4.0 – 6.0
PLA	212	4
PPS	275	3.0 – 6.0
PS-GPPS	180	2.0 – 4.0
PS-HIPS	160 – 180	2.0 – 4.0
PSU	275	4
PUR	N/A	N/A
PVC	150	2.0 – 4.0
PVDF	302	1
SAN	160 – 180	2.0 – 4.0
SPS	176	2.0 – 4.0

some cases discoloration can be seen in the plastic pellets. It is also possible that the plastic pellets may soften and stick together in the dryer. The pellets affected the most are those near the exit, or bottom, of the hopper, which take the weight of the pellets above them, causing bridging and making the situation worse. Higher temperatures can also lead to the loss of the low molecular weight additives in the base polymer.

In case of both hygroscopic and non-hygroscopic polymers, there can be water condensation on the pellet surface. If the pellet is transported from a relatively cold environment, such as a silo on the outside of the building, into a warmer and possibly more humid molding environment, water can condense on the surface of the pellets. This is similar to the water condensation we see on the outside of the glass of a cold drink on humid days. Condensation can also occur at the feed throat of the molding machine which is typically kept cool by cold water circulation in order to avoid bridging at the feed throat. If the area is not cooled, the plastic that is stationary between the end of screw recovery and before the mold opens melts to form a solid mass and prevents the plastic from being conveyed into the molding barrel. Non-hygroscopic materials are therefore surface dried at low temperatures to remove this moisture before processing. Exposing the plastic to room temperature or low temperature dry air for a short period of time is sufficient to dry off the moisture. The drying time can be as low as half an hour on low humidity days.

Additives can also be hygroscopic in nature. For example, certain calcium-based fillers can absorb moisture. Any plastics filled with these compounds must therefore be dried before processing regardless whether the base plastic is hygroscopic or not.

4.3.2 Relative Humidity and Dewpoint

To extract the moisture out of the plastic, the air used for drying should be as dry as possible. When the plastic with moisture is exposed to the dry air, the system is looking to achieve equilibrium and therefore the moisture is continually extracted. The dryness of the air can be expressed with the help of two terms: the relative humidity and the dew point. Relative humidity is the percentage of moisture the sample of air holds compared to the amount of moisture it could hold when it is saturated. The saturation level changes with temperature. The lower the temperature, the lower is the maximum amount of moisture it can hold.

Dew point indicates the amount of moisture in the sample of air. The higher the dew point, the higher is the amount of moisture in the air. The dew point is defined as the temperature to which the given air sample will have to cool in order to reach 100 % relative humidity or complete saturation. Therefore, a lower dew point indicates a lower amount of moisture in the air. A temperature of –40 °C (–40 °F) equates to very low amount of moisture (less then 0.4 %) in the sample of the air. This temperature is therefore taken as the target dew point temperature to be achieved for the air that is supplied to the dryers.

Let us put this into perspective by considering an example. The dryer temperature is set to 100 °C. This means that the air being supplied to the dryer is at 100 °C. At this temperature, the air may still contain moisture, depending on its relative humidity. This moisture will prevent the drying of the plastic to the required levels. To reduce this moisture, the air must be dried. This is usually done by using desiccant beds that absorb the moisture from the air as it is passed through the desiccant. As the amount of moisture in the air is reduced, the temperature at which it will condense, or its dew point, also reduces. Lower dew point temperatures indicate that the amount of moisture being supplied to the dryer is also lower. Therefore, measuring the dew point will indicate the dryness of the air. If the dew point is around –40 °C, the quality of the air is acceptable to dry the plastic. The final moisture level in the air is a result of the temperature, relative humidity, and the dew point of the air.

4.3.3 Air Flow Rate

The hopper used to dry the material is supplied with air that has a low dew point, is dry, and has been heated to the recommended temperature. As the plastic dries, the moisture migrates to the surface and should be continuously carried out of the system. This requires a sufficient amount of flow of the air through the system. Therefore the flow rate of the air through the system is important.

4.4 Equipment for Drying Plastics

Several types of dryers are available for drying plastics. The classification is based on the technology or other features, such as location of the dryer.

4.4.1 Oven Dryers

During the early years of molding, when processors realized the need for drying, ovens similar to baking ovens were used. The plastic was spread out on large trays and put into the ovens to dry. The large trays helped to spread out the plastic and increased the area of exposure to the heat. The plastic would not dry evenly if the layer of plastic was too thick. Handling of the hot trays and transporting the material to the hopper of the machine was not easy and often required two people. Smaller ovens are still in use in some R&D facilities and production facilities that process small amounts of material. One of the additions to these ovens is the use of a vacuum pump. The vacuum decreases the boiling point of any liquids and at the same time facilitates the removal of the moisture from the chamber speeding the drying process.

4.4.2 Hot Air Dryers

Hot air dryers supply hot air to the plastic at the bottom of the dryer and as the air passes through the plastic, it picks up the moisture from the surface of the pellets and transports it out of the system. The heat also helps to drive the moisture to the surface of the pellet. The supplied air is picked up from the atmosphere, heated and pumped into the dryer. These systems are best suited for surface drying of non-hygroscopic materials such as olefins or for materials that do not require moisture levels to be very low before processing.

4.4.3 Desiccant Dryers

These dryers are similar to hot air dryers except for the fact that the air first passes over a desiccant bed that adsorbs almost all of the moisture from the air. This dry air is then heated and supplied to the hopper. Since the relative humidity is now low, the air can pick up more of the moisture from the plastic, helping to achieve the lower moisture levels desired for materials such as nylons. As the moisture is adsorbed by the desiccant bed, the desiccant will eventually become saturated with the absorbed moisture. As the saturation level of the desiccant increases, its ability to absorb moisture decreases, making the drying process less efficient. To correct this, there is a regeneration cycle during which the desiccant is dried and then recycled back into the system. Desiccant dryers are widely used in most modern-day facilities because of their versatility.

4.4.4 Classifications Based on the Location of the Dryer

Dryers can be located next to the molding machine and a hose can be used to supply dried material from the dryer to the molding machine. These are conventional dryers that are most commonly used in small to medium size companies in which a variety of materials are processed and mold changes are frequent. The dryers are mobile and can be moved from one machine to another. In companies processing a mix of hygroscopic and non-hygroscopic materials it is not required to invest in the same number of dryers as the number of molding machines. The dryers can be moved where needed. Hopper dryers on the other hand are dryers that are mounted directly on the molding machine such that the dry material is fed straight into the feed throat of the molding machine. This is a good solution for extremely hygroscopic materials such as some polyurethanes that can pick up moisture during the transportation from the conventional dryer to the molding machine through the transport hose. The disadvantage is that these dryers are attached directly to the molding machine and are not easily moved to different molding machines on short notice. This type of dryer, for all practical purposes, is considered dedicated to the machine.

Central dryers are very large dryers that can be installed in companies that process large quantities of the same material. The material is dried at a central location and is then transported to the molding machine for processing.

4.5 Determination of the Amount of Moisture

It is critical to be able to accurately determine the amount of moisture in the plastic to be melt processed. Therefore, an accurate measurement system is necessary. There are various direct and indirect measurement methods, most of which suffer from a basic problem. When trying to extract the moisture for analysis and measurement, other volatiles, additives, and residual polymerization compounds including monomers are released. These can therefore mask the actual amount of moisture in the plastic and give false readings. There are ways to prevent this but they would not be practical in a production environment. Some moisture analysis methods are discussed in the following.

4.5.1 The Glass Slide Technique (TVI Test)

This test was developed by GE Plastics (now Sabic Innovative Plastics) and is called the Thomasetti Volatile Indicator Test or the TVI test. It involves placing four to five plastic pellets on a glass slide that is heated on a hot plate. With the help of another slide, the pellets are pressed together and sandwiched between the two slides. Once the pellets are flattened out, the slides are removed and allowed to cool. Any moisture in the pellets shows up as bubbles in the slide. Sometimes it is helpful to use a microscope to view the bubbles. This is an easy way to determine the presence of moisture in the sample, although the slides and the hot plate must be handled very carefully to avoid safety issues. This test does not provide a numeric value of the amount of moisture present in the plastic and, as in other tests, volatiles from degradation or an additive cannot be identified. The TVI Test is not a commonly used test in industry.

4.5.2 The Karl-Fischer Titration Method

This type of analysis will produce the most accurate results. However, because of the time required to conduct this test, the additional materials and equipment required, the test is also not commonly used. The basic underlying principle of this method is that a small amount of electricity will be generated with the chemicals involved in the test due to the reaction of the moisture in the plastic. The plastic is heated to a higher temperature to release the moisture. The amount of electricity generated is directly related to the amount of moisture present in the system. Accurate measurement of the electricity generated provides the amount of moisture present in the plastic. Figure 4.8 shows a setup for a Karl Fisher titration system. A disadvantage of this system is that at elevated temperatures, water can be produced in some plastics. This may be caused by degradation or by melt polymerization, a phenomenon explained in Section 4.5.2. Therefore, for these plastics the newly formed water can skew the results and provide false readings.

Figure 4.8 A Karl Fischer setup to measure moisture in a sample (Courtesy: Denver Instruments)

4.5.3 Electronic Moisture Analyzer

With the advances in technologies and with better understanding of the moisture absorption process, electronic moisture analyzers have become common in many production facilities. The biggest advantage they provide is their simplicity of use without any prior knowledge or experience in plastics or moisture determination. The test takes less than approx. ten min-

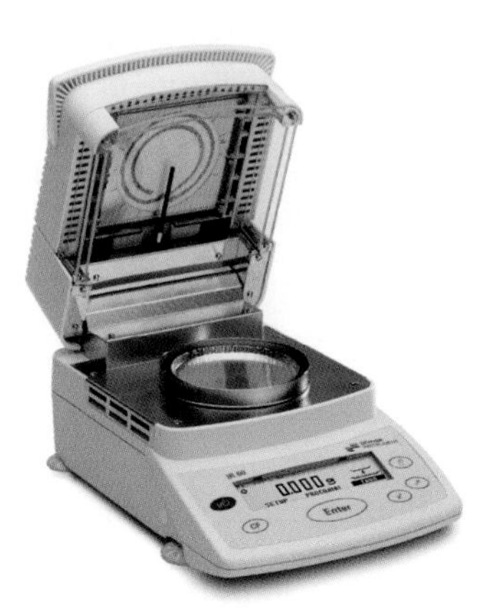

Figure 4.9 Electronic moisture analyzer (Courtesy: Denver Instruments)

utes and provides a print out of the moisture level in the plastic. The instrument can also be hooked up to a PC to collect and record test data over time. These analyzers work on the principle of weight loss when the plastic is heated and the water leaves the system. In this test, a small amount of plastic is placed on a pan that is attached to a micro-scale. The weight of the sample is recorded and the sample is heated to a desired predetermined temperature for each plastic. As the water leaves the plastic, the weight drops and once the weight stays constant, it is assumed that the plastic has lost all its moisture. The final weight is recorded and the percentage weight loss is calculated, providing the moisture content of the plastic. The disadvantage of this system is that it does not take any other volatiles that may be lost during the heating process into consideration, which could influence the moisture content results. Figure 4.9 shows a picture of a moisture analyzer. These types of analyzers are widely used because of the simplicity of use and the sufficiently reliable data they produce.

4.5.4 Measurement of the Dew Point

A dew point meter helps to measure the dew point of the air that is supplied to the dryer. Although this method does not actually measure the dryness of the resin, it helps in assuring that the dryer is fed with dry air. A dew point of –40 °C (–40 °F) is a good indication that the amount of moisture in the air is at very low acceptable levels.

4.6 'Overdrying' or Overexposure to Drying Temperatures

Most polymers and plastics are blended with low molecular weight additives, such as heat stabilizers, processing aids, and other specialty additives. The additives are used to enhance the plastics' properties for specific applications and/or to reduce their cost. Such a blended polymer is called a resin. All plastics used for melt processing are considered resins because of these essential additives. Additives include plasticizers, lubricants, flame retardants, heat stabilizers, colorants, blowing agents, and biocides among others. Most of them are added to the polymers in small percentages and are low molecular weight compounds or oligomers.

The drying of the plastic must be controlled in order to not exceed the maximum recommended drying times. On the production floor this important rule is very easily overlooked. It is typical in the molding industry to first load the resin into the dryer and then to complete the mold change in order to ensure that the resin is being dried during the setup operation and no time is being wasted. This is a very efficient procedure. However, if the mold setup was not completed in time or the first shots from the mold were not acceptable and the mold had to be pulled out for maintenance, the plastic that is left in the dryer is now being subjected to additional drying time at elevated temperatures. The overexposure may lead to a potential loss of the low molecular weight additives in the plastic. These additives are usually not as heat resistant as the polymers, especially after prolonged drying times. Another scenario in which over-drying can be a potential problem is when the dryer is oversized for the mold in the machine. In an oversized dryer the residence time of the plastic is longer than the recommended maximum drying time of the plastic. For example, if the hourly usage of material is 10 lbs and the maximum recommended drying time is 8 hours, then, if the capacity of the dryer is more than 80 lbs, the plastic that is on top or that is loaded last, will be subjected to a drying time of more than 8 hours before it is actually used. This must be avoided.

Case Study

A study [1] exemplifies this for two particular materials, a polyester and a nylon. Here, the polyester was a 30% glass filled polybutylene therpthalate (PBT) and the nylon was a 15% glass filled nylon 66. Both resins were dried for varying times and the impact of the overdrying was studied. Parts were molded from an existing production mold. For each material, the process was kept unchanged during the molding of the resins dried for the various times. The results were material specific and the effect of overdrying was different for both materials, as was confirmed by thermal analysis and mechanical tests. In a thermogravimetric analysis (TGA) the plastic sample is continuously heated, which causes the polymer to burn off at a given temperature which depends on the base plastic and its additives. Rheology studies are performed to study the viscosity of the plastic and to determine the effect of drying. The shear sweep study provided the viscosity versus shear rate plot and the thermal degradation study was done by plotting the viscosity versus time at a given shear rate. As the plastic stayed in the barrel, the plastic degraded and a change in the viscosity was recorded. The observations and the results are discussed in the following.

Results for PBT

The resin and the molded parts were tested using the procedures mentioned above. During the molding process it was seen that as the drying time increased, flash on the parts increased, indicating a drop in melt viscosity. However, there was no significant change in the actual characteristic values, such as fill time, cushion value, screw recovery time, and others. The TGA data is shown in Fig. 4.10.

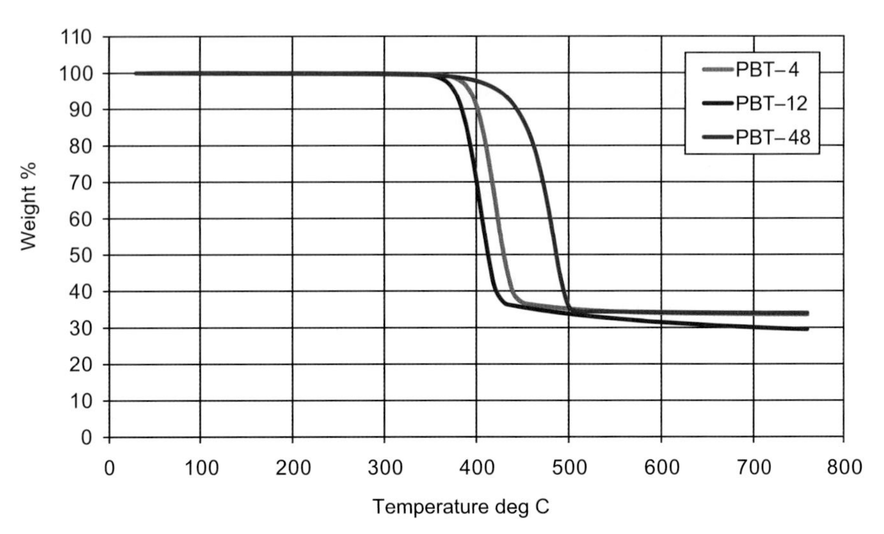

Figure 4.10 TGA graphs of PBT dried for 4, 12, and 48 hours [1]

The final residue left behind was about 33 % of the initial weight. Since the resin was a 30 % glass filled material, most of this residue must have been the glass. To better understand the results, $T_{1/2}$ is defined as the temperature at which 50 % of the weight loss occurs. Considering the percentage of the residue remaining, 50 % of the weight loss occurs at approx. 420 °C for PBT-4 and at approx. 475 °C for the PBT-48. This indicates that for the shorter drying times, a weight loss occurred at lower temperatures suggesting there must be a component in the resin that decomposed at lower temperature leading to an earlier loss in weight. This additive was probably decomposed and was taken out of the dryer during the lengthy drying process and therefore the PBT–48 had a $T_{1/2}$ that was almost 55 °C higher than the $T_{1/2}$ for PBT-4. Looking closely at the numerical data and analyzing the initial weight loss in the PBT, it can be seen that the slope of the curve is higher for the PBT-4, indicating sudden weight loss similar to those seen in low molecular weight compounds and oligomers. After the initial differences in the slopes of the curves, the lines then seem to run parallel, suggesting the decomposition of the base resin.

Shear sweep data obtained by capillary rheometry is shown in Fig. 4.11. The data did not show any difference in melt rheology. This could suggest that the additive lost during the drying process was not a processing aid intended to lower viscosity for ease of processing. No other conclusions could be drawn.

Thermal degradation data also obtained using capillary rheometry is shown in Fig. 4.12. The data showed a difference between the two resins only at residence times of 9 minutes and above. PBT-48 showed a lower viscosity compared to PBT-4. This could have been caused by the degradation of the base resin, lowering its molecular weight and thereby lowering its viscosity. It could also be an indication that the component lost during the excessive drying process was a heat stabilizer. It is also interesting to note that the viscosity curve almost flattens out past 9 minutes, possibly suggesting that the polymer is completely degraded and that it is the glass fibers now being carried by the degraded resin and contributing to the viscosity.

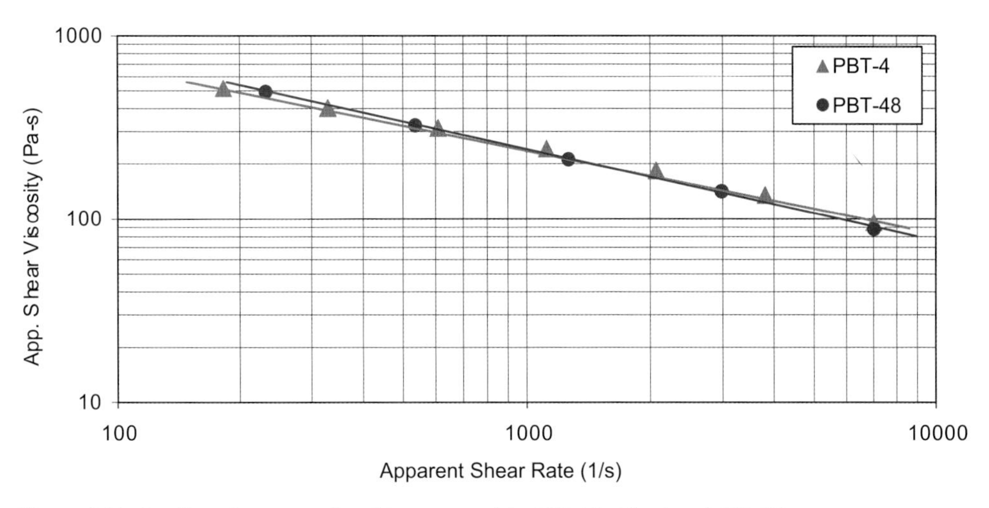

Figure 4.11 Capillary rheometry data (shear sweep) for PBT dried for 4 and 48 h [1]

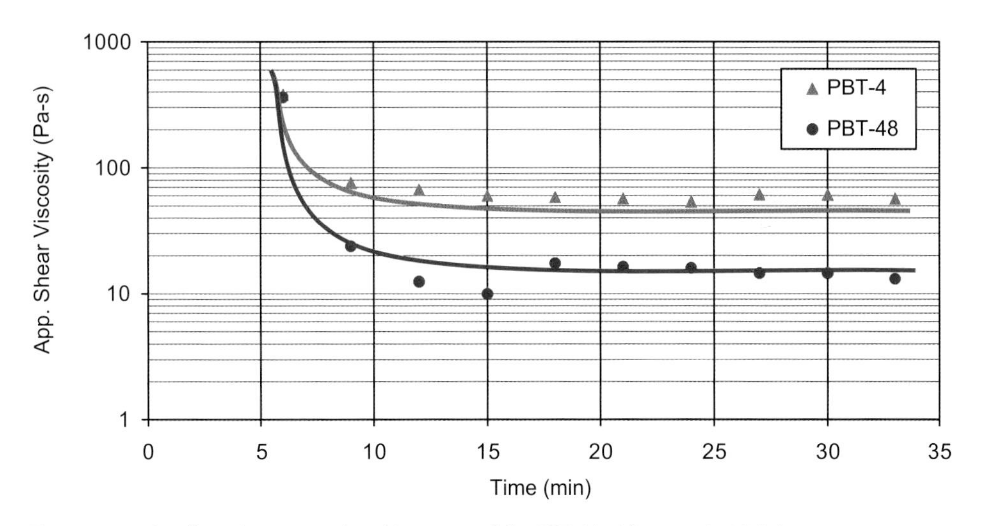

Figure 4.12 Capillary rheometry data (time sweep) for PBT dried for 4 and 48 h [1]

PBT parts molded from this resin after excessive residence times in the dryer crumbled into pieces when a small amount of force was applied. The drop impact testing of the parts demonstrated a clear difference between PBT-4 and PBT-48. This data is shown in Fig. 4.13. As can be seen, the mean failure energy decreased with the increase in drying time. The parts tended to become more brittle. This could be the effect of either the loss of an additive, such as an impact modifier, and/or the degradation of the resin. There was a sharp drop in the mean failure energy between 12 and 36 hours of drying time. Before and after these times the curve stayed relatively flat. Looking at this data, drying times should be limited to 12 hours to retain the material properties.

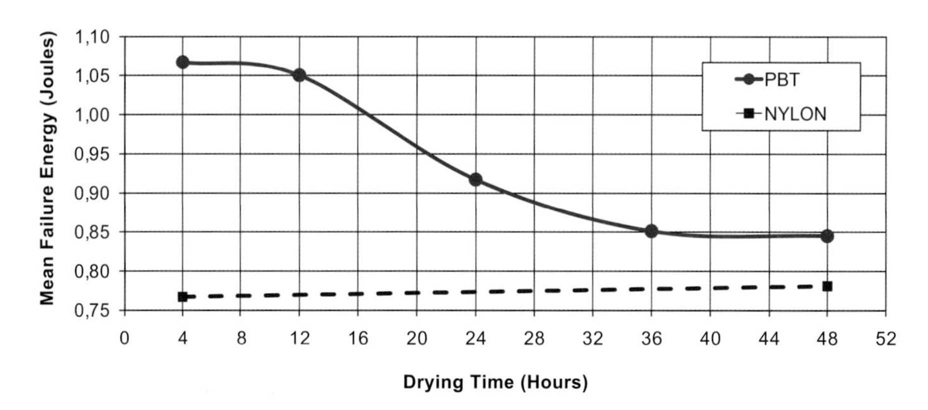

Figure 4.13 Mean failure energy of parts molded with PBT and nylon dried for varying drying times [1]

Results for Nylon

In the case of nylon, studies were conducted for drying times of 4 hours and 48 hours. The resin and the molded parts were tested with the procedures mentioned earlier and the TGA data is shown in Fig. 4.14. Here, a 15 % glass filled resin was used. The residue left behind was approximately 17 % of the original weight. We can again be certain that most of the residue was the glass left behind. The $T_{1/2}$ values for both nylon-4 and nylon-48 seemed to be approx. 470 °C. The weight loss also started around the same time and imitated each other. We can therefore infer that there was no significant difference between the TGA curves for the regularly dried and the overdried resin. There was no noticeable loss of any additive.

The capillary rheometry-shear sweep data is shown in Fig. 4.15. It is interesting to note that the viscosity for the nylon-48 was higher than that for nylon-4. Khanna, *et al.* [2] have noticed a similar increase in the viscosity of nylons with increase in drying temperature and

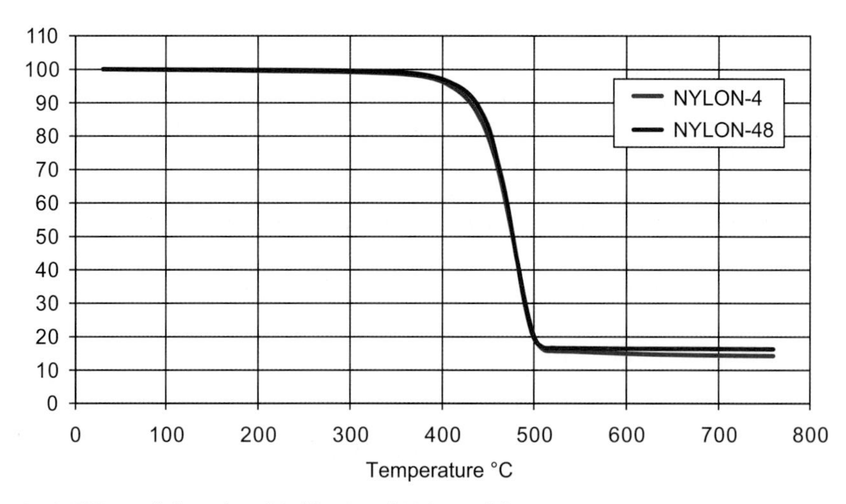

Figure 4.14 TGA graph for nylon dried for 4 and 48 hours [1]

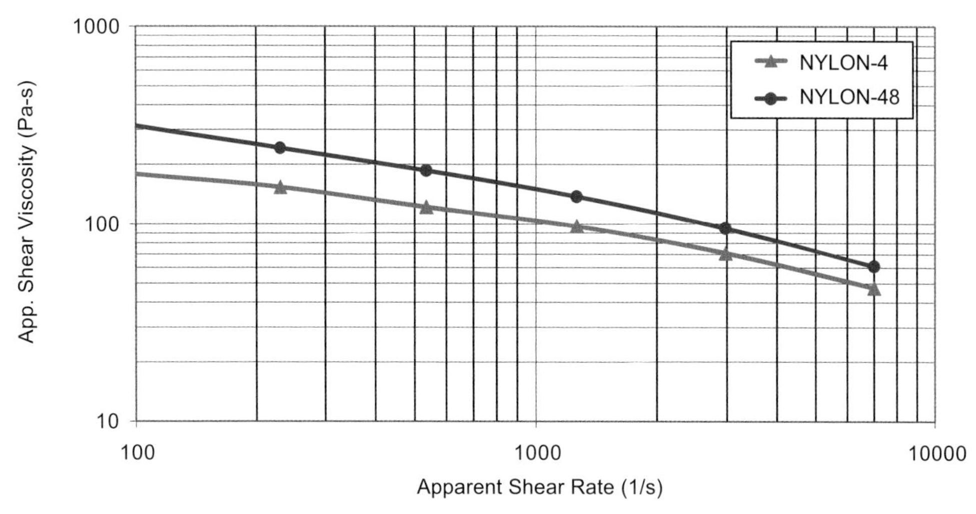

Figure 4.15 Capillary rheometry data (shear sweep) for nylon dried for 4 and 48 h [1]

time held constant. Similar studies by Pezzin and Gechele [3] showed that the melt viscosity increased with time for lower moisture contents. Khanna hypothesizes that the increase in the viscosity could be explained by the following two reasons:

- The moisture in the nylon acts as a plasticizer for the melt, reducing its viscosity. With higher drying temperatures, loss of moisture leads to an increase in viscosity.
- The melt equilibrium is represented by a characteristic equilibrium constant, K_{cond}. If excess water is added to the melt in equilibrium, the reaction will go in the direction of the reactants (i.e., degradation), while the reverse (i.e., polycondensation) should occur if water is removed from the equilibrium melt.

In the present study, the drying temperature was held constant and the drying time was varied. Considering the theories put forth by Khanna, and the similar results obtained here, the net result of the increase in drying time must be the same as the increase in drying temperature. The loss of moisture between drying times of 4 hours and 48 hours must have contributed to the increase in viscosity of the melt. Evidence of this was also seen during the molding of the parts. A feature that was 48 mm long with an average height of 3.5 mm and a width of 6 mm was almost completely filled (99 %) during the injection phase of the molding cycle when molding with nylon-4. When nylon-48 was used, this feature was only filled to approx. 62 % of the original flow length. Photographs of this are shown in Fig. 4.16.

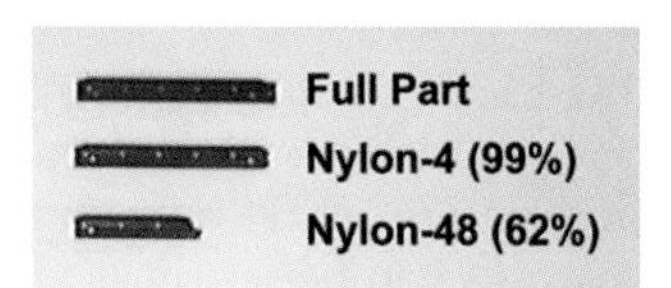

Figure 4.16 Complete parts and 'injection only' parts molded with nylon dried for 4 and 48 h [1]

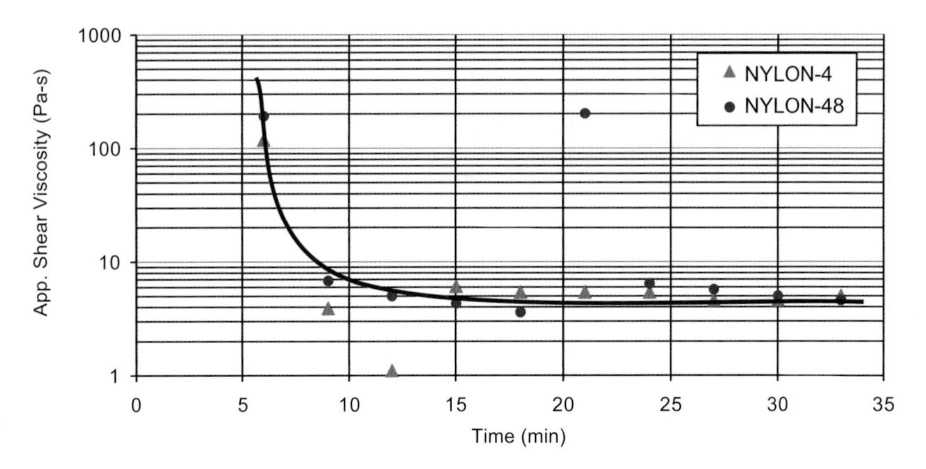

Figure 4.17 Capillary rheometry data (time sweep) for nylon dried for 4 and 48 h [1]

Thermal degradation data using capillary rheometry are shown in Fig. 4.17. The two resins do not seem to differ in viscosity for residence times of 12 minutes and beyond. With a residence time of 6 minutes, nylon-48 has a higher viscosity compared to nylon-4. The same reasoning as in the above section can be applied here. Once past the 6 minute interval, the resin probably begins to degrade and the final viscosity of the two resins matches.

Drop impact testing of the parts demonstrated no significant difference between nylon-4 and nylon-48 as shown as a dotted line in Fig. 4.13. The average mean failure energy for the two samples was 0.774 Joules. Long drying times did not seem to affect the drop impact strength of the resin. It is evident from the above discussion that it is important to control the drying process of the resin. Overdrying can result in a loss of physical properties as seen in the case of PBT, or lead to the increase in the viscosity as in the case of nylon. For nylons, because the water in the plastic has an effect on the viscosity of the plastic, the water can be thought of as a viscosity regulator. The viscosity changes will have an impact on the flow properties and the associated features, such as the weld line strength of parts. Other properties affected by flow, such as surface finish and polymer/filler ratio could also be impacted by long drying times. Secondary operations, such as ultrasonic welding and joining, require a certain amount of polymer to be present on the surface in order to achieve successful bonding. Therefore, a minimum level of moisture is required for nylons because water is a viscosity regulator. This level is usually around 0.015 %. However, this is something each molder must determine and maintain individually.

4.7 Cautions

Drying of some resins such as PBT has a cumulative effect. The drying process removes the low molecular weight additives for good and they do not return if the drying is stopped and the resin returned to the shelf for the next run. However, in the case of nylons, the moisture can be absorbed back and the viscosity can be re-regulated. Since each resin and its additives is unique, customized experimentation is best suited to determine the type of control required for each individual process. The results above must not be taken as any sort of standard results. Experimentation is time consuming and expensive and therefore it is best to keep it simple and avoid overdrying.

4.8 Prevention of Overexposure to Longer Drying Times

On the shop floor there are several efficient ways to prevent overdrying:

Turning down the dryer temperature: If molding is not ready to be started after the material is dry, turn down the dryer temperature to about 25 °C (80 °F), but keep the dryer running. Supplying the hopper with low temperature dry air will keep the moisture out and will not have detrimental effects on the resin.

Sizing the hopper: Size the hopper dryer such that the residence time of the plastic stays between the minimum and maximum recommended drying times of the plastic. When the hopper is larger than required this may be difficult. In this case, retrofit the dryers with a level sensor and adjust the level to maintain the required amount of material.

The material required for the run must be calculated and only the required amount of material should be dried. If the machine is going to be down for an extended period of time after the drying has taken place, the dryer must be shut off as soon as possible.

4.9 Overdrying Controller

Figure 4.18 shows this logic applied to a design of a controller, while Fig. 4.19 depicts a concept for a programmable controller that would control the drying process of the resin.

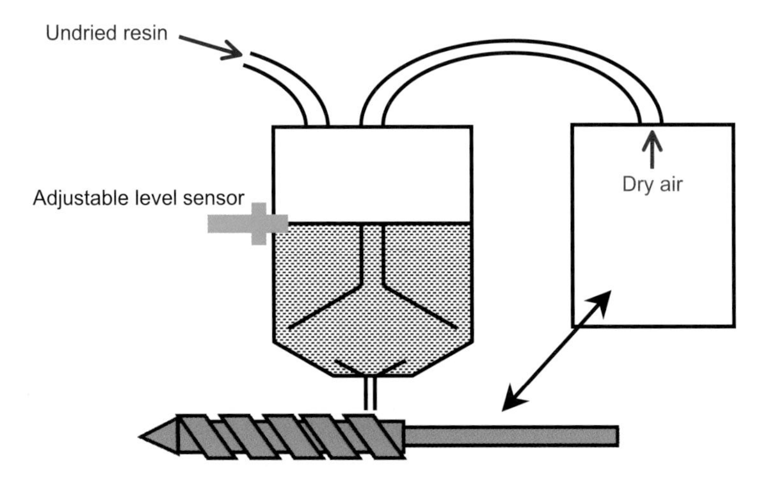

Figure 4.18 Controller logic to control residence time in the dryer [1]

Initially, the controller would set the drying time and temperature. Based on shot weight of the mold, a resin level sensor would be set such that the residence time of the resin in the hopper exceeds the recommended drying time by an hour to two hours. Once the set drying time has elapsed, the controller looks for a signal from the machine indicating that the machine is operational and molding parts. Such a signal can be picked up from screw rotation or mold open/close. If the controller receives this signal it will maintain the dryer temperature. If it does not receive the signal, indicating that the machine is not operational, the controller will start to drop the drying temperature by a preset value, for example 10 °C. If in a preset time, for example 15 minutes, the controller does not get the machine operation signal, it will further drop the temperature down by another step. It will do so until it reaches a temperature of 25 °C. At this time, dry air will be circulating in the hopper and keep the resin dry. As soon as the machine is ready for operation and the controller gets this signal, the temperature of the hopper will begin to rise to the desired drying temperature in preset steps of temperature and time. If normal molding operation is interrupted, the controller will follow the same logic described above and drop the dryer temperature. Such a control mechanism will ensure that the resin will never experience excessive drying times and prevent overdrying.

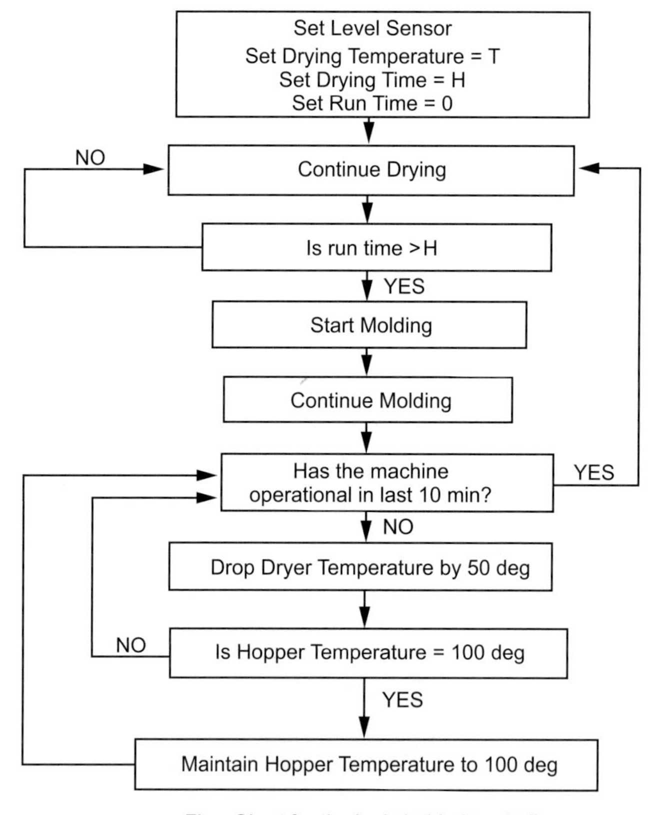

Figure 4.19 Flowchart for the logic behind controlling the drying process [1]

References

1. Kulkarni, S.M., *SPE ANTEC Tech Papers* (2003) p. 736
2. Khanna Y., et al., *Polymer Engineering and Science*, Vol. 36 (13), p. 1745, (July 15, 1996)
3. Pezzin G., and Gechele G., *J. Appl. Polymer Sci.*, Vol 8, p. 2195, (1964)

Suggested Further Reading

1. Brydson, J.A., Plastics Materials (1999) Butterworth Heinemann Ltd, Oxford
2. Harper, C.A., Modern Plastics Handbook (2000) McGraw Hill, New York, NY
3. Deanin, R.D., Polymer Structure, Properties and Applications (1972) Cahners, Boston, MA
4. Odian, G., Principles of Polymerization (1991) Wiley Interscience, NY
5. Shah, V., Handbook of Plastics Testing and Failure Analysis (2007) Wiley Interscience, NY

5 Common Plastic Materials and Additives

Injection molding processors are mainly concerned with the way the plastic flows and with the temperature at which the plastic can be processed. Flow characteristics of all plastics are very similar. They all follow non-Newtonian behavior and therefore exhibit shear thinning effects. A pattern from a flow simulation program for two different plastics will look very similar, because plastic flow is always laminar. The pressures and temperature distributions may be different, but the flow patterns are similar. All plastics have their unique processing temperature range, both for the mold temperature and the melt temperature. Even with these commonalities, it is still important to understand the different types of plastics and the additives that are incorporated into them. With this knowledge processors will be able to better understand the materials they work with and to take the necessary precautions during their processing. For example, while polyethylenes can be injected at very high speeds, one must be very careful with PVC, because it tends to degrade at higher speeds. This chapter will introduce the base materials, the additives, and the reason for their incorporation. Only the most common polymers by volume and the additives used in injection molding will be discussed.

5.1 Classification of Polymers

Polymers can be classified in a number of ways. In the field of molding the term plastic is most commonly used. We use the term polymer when we describe the basic nature of the molecules and their properties. Plastics are those polymers that can withstand a moderate to significant amount of force before showing any significant deformation. Polymers that are deformed under light loads are called elastomers. Technically, the difference can be seen clearly in the stress-strain graphs in Figure 5.1. In the following, the term plastic will be used to describe all molding materials and the term polymer will be used when an intrinsic property needs to be referenced.

The following classifications of polymers are also used:

Thermoplastics: They can be repeatedly heated, melted, and processed into useful products. Example: ABS

Thermosets: Once these polymers are processed, they form a chemical network and all the molecules now are crosslinked. It is impossible to re-melt the polymer in this state because there is more energy required to separate the crosslinked bonds than the energy required to break the main chains. This results in a total destruction of the polymer. Example: Liquid silicone rubbers (LSRs).

Organic Polymers: These contain carbon atoms in their backbone and are derived mainly from organic materials from nature. Example: Polyethylene

Inorganic Polymers: These contain atoms other than carbon in their backbone.
Example: Polysilanes

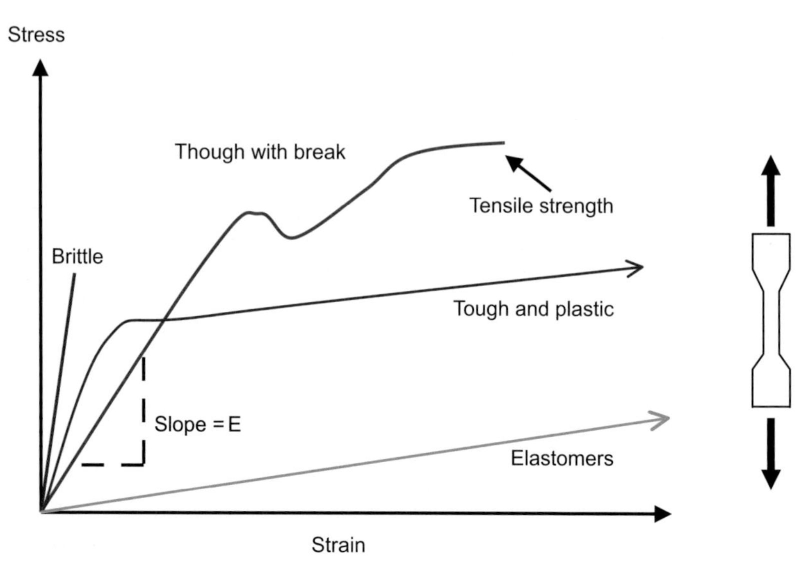

Figure 5.1 Tensile properties for different types of plastics

Elastomers (TPE): These are similar to thermoplastics except they are soft and rubbery at room temperature. Their glass transition temperatures are below room temperature.

Homopolymers: These are polymerized from only one type of monomer.
Example: Polyethylene

Copolymers: These are manufactured from two or more monomers.
Example: ABS – polymerized from acrylonitrile, butadiene, styrene.

Alloys: These are polymers physically mixed with each other and there is no chemical interaction between the different polymers. Example: PC-ABS alloys are common in the appliance industry.

5.2 Commercially Important Plastics

Some of the commercially important plastics that are injection molded are described in the following. All descriptions relate to unfilled plastics without any additives.

5.2.1 Polyolefins

Polyolefins are the most basic hydrocarbons and contain carbon and hydrogen atoms only. Polyethylenes and polypropylenes are the most widely injection molded polyolefins.

Polyethylene: Polyethylene is one of the widest used materials because of its excellent electrical insulation properties, good chemical resistance, impact properties, and low cost. The properties of polyethylene (also known as polyethene) can be tailored to meet certain application requirements. The tailoring changes the density of the polymer which allows to classify the different polyethylenes. Low density polyethylene (LDPE) has a density of $0.91-0.92\,g/cm^3$, medium density polyethylene (MDPE) has a density of $0.93-0.94\,g/cm^3$ and high density polyethylene (HDPE) has a density of $0.95-0.96\,g/cm^3$. Linear low density polyethylene (LLDPE) is technically a copolymer and is designed for better impact properties and improved flexibility. It is mainly used for films. The density of LLDPE is around $0.91\,g/cm^3$. To improve some of the mechanical properties and still retain the other properties of polyethylene, its molecular weight is increased. This product is called ultra high molecular weight polyethylene (UHMWPE) with a density of $0.92-0.93\,g/cm^3$. UHMWPE is difficult to mold because of its high melt viscosity. The weight average molecular weight of UHMWPE can be in the range of $1-6 \times 10^6$, whereas for other ethylenes it is around 50,000 to 300,000. The disadvantage of polyethylenes is that they exhibit high thermal expansion, poor weathering resistance, and low heat deflection temperatures. Although the mechanical properties of polyethylene are low, with the use of reinforcing fillers such as glass fibers these properties can be enhanced. Polyethylenes can also be crosslinked to improve their properties.

Polypropylene: Both polypropylene and polyethylene have similar structures and similar properties. Most commercially available polypropylenes are isotactic polypropylenes with densities of $0.90\,g/cm^3$. They exhibit good electrical properties, good environmental stress cracking resistance, and good heat resistance. Polypropylene dominates in the area of thin-wall molding which requires good flex properties. CD covers with living hinges are examples of these applications that can stand continued flexing. The biggest disadvantage of polypropylene is its low temperature flexibility or impact resistance. As the temperature reaches 0 °C, the polymer becomes brittle and often cracks. This issue is often resolved by either incorporating additives or copolymerizing polypropylene with a small percentage of polyethylene.

Other olefins such as cyclic olefins are used in molding but not to the extent as polyethylenes and polypropylenes.

5.2.2 Polymers from Acrylonitrile, Butadiene, Styrene, and Acrylate

Monomers and polymers of acrylonitrile, butadiene, styrene and acrylates are compatible with each other and produce a variety of useful polymers and plastics. Each of these can be polymerized by themselves or with other monomers. Each of the monomers provides a unique property to the final polymer. For example, butadiene is a polymer with a low T_g value and therefore provides impact resistance; styrene adds luster to the product and helps to improve the processability; acrylonitrile adds heat and chemical resistance; acrylic adds weathering resistance. ABS is one of the most commonly used materials. Depending on the end use requirement, it can be produced with varying percentages of the monomers. An absence of styrene produces nitrile rubber, an absence of butadiene will produce

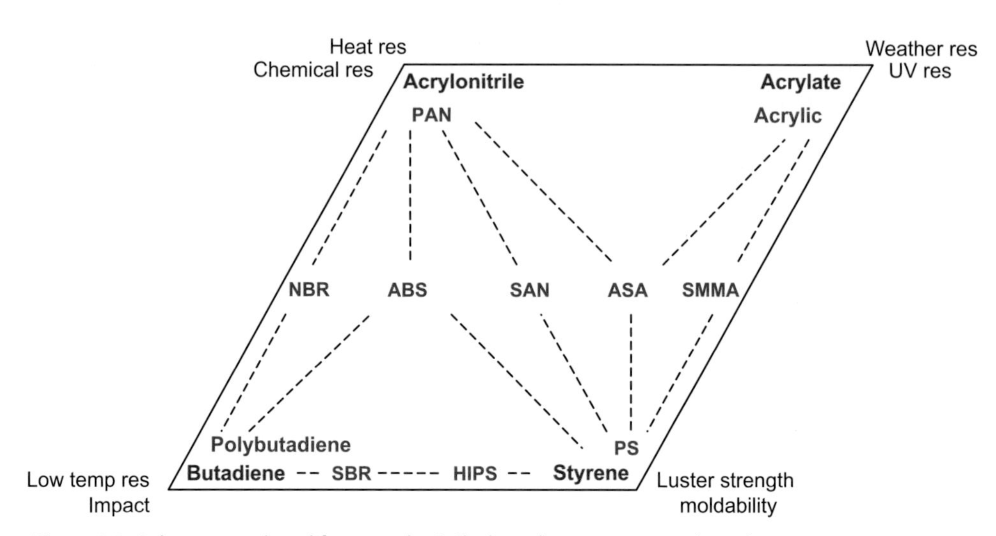

Figure 5.2 Polymers produced from acrylonitrile, butadiene, styrene and acrylates

styrene acrylate resins (SAN), and an absence of acrylate will produce styrene butadiene rubber (SBR) or impact modified styrene (HIPS). SBR is butadiene with a small percentage of polystyrene and HIPS is polystyrene with a small amount of butadiene. By themselves, the polymerization products are polyacrylonitrile, polybutadiene, and polystyrene. Similar combinations with acrylate monomer yield different polymers as was shown in Fig 5.2. Because of the endless possible combinations of the monomers, the properties of the end polymer can be tailored. Not only the percentages of the monomers play an important role, but the method of preparation can also affect the properties. The polymers can be copolymerized from the monomers, grafted onto a base polymer, or the individual polymers can be physically blended with each other. These polymers can be further blended with other polymers to create other customized polymers. A blend of polycarbonate (PC) and ABS called PC/ABS is widely used in thin-wall molding, such as cell phone cases.

5.2.3 Polyamides (PA)

Most polyamides used for molding are commonly called nylons, although not all PAs are nylons. Nylons are crystalline materials; they are manufactured by polycondensation reactions usually producing water as the by-product. Nylons produced from dibasic acids and diamines are identified by a numbering system that indicates the number of carbon atoms in the base monomers. A nylon 6/6 has six carbon atoms in the dibasic acid and six carbon atoms in the diamine. A nylon 4/12 has four carbon atoms in the dibasic acid and twelve carbon atoms in the diamine. For example, nylon 6/6 is made from hexamethylene diamine and adipic acid that each contain six carbon atoms. Nylons made from a single monomer are identified by just one number. For example nylon 6 is made by the ring opening polymerization of caprolactam that contains six carbon atoms. Commercially, nylon 6 and nylon 6/6 are widely used in injection molding. Nylons are tough and offer good impact and chemical resistance even at moderately elevated temperatures. This is the reason they are used in a lot of automotive applicatioons, such as under the hood. Gears, cams, and bearings are also products made from nylons. The biggest disadvantage of nylons is due to the fact that they are strong hydrophilic materials, i.e., they absorb a lot of moisture. The presence of moisture drastically affects their properties, such as dimensional stability and electrical properties. Most data sheets for nylons are therefore available for 'dry' and 'conditioned' nylons. The properties of nylons are also affected by the crystallinity of the polymers. Table 5.1 shows the relative difference in properties between nylon 6 and nylon 6/6.

Table 5.1 Relative Differences in Properties of Nylon 6 and Nylon 6/6

Property	Nylon 6	Nylon 6/6
Crystallinity	Lower	Higher
Heat resistance	Lower	Higher
Chemical resistance	Same	Same
Impact strength	Higher	Lower
Wear resistance	Lower	Higher
Processibility	Lower melt and wider window	Higher melt and narrower window

Processing of nylon if fairly simple because of its favorable flow properties. Nylon must be dried to the right moisture level before processing. Some amount of moisture in the plastic acts as a viscosity reducer and regulator and therefore exposure of nylon to excessive drying times should be avoided. Drying of nylon is discussed on Chapter 4. If nylon is not dried prior to molding, it can degrade in the barrel due to hydrolysis causing a drop in properties. Surface defects such as splay is common with undried resin. Drooling from the nozzle tip between mold open and mold close is another issue because the water reduces the viscosity of the plastic and makes it drool out of the nozzle tip after the shot builds up. Because of the high T_m and T_g, cycle times can be short.

5.2.4 Polystyrenes (PS)

Polystyrene is polymerized from styrene and is also called poly(vinyl benzene). It can be either atactic or syndiotactic, as described on Chapter 2. Most commercially used polystyrene is atactic with good optical clarity because of its amorphous nature. The T_g of the polymer is high, resulting in good dimensional stability. Solvent resistance, weathering resistance, and impact strength of polystyrenes are low. Impact strength can be improved by incorporating a small amount of rubber, such as butadiene, creating a high impact polystyrene, commercially called HIPS; however, this affects the clarity of the final product. Another useful property of polystyrene is its ease to be processed into foam. Commercially available styrofoam is polystyrene mixed with a blowing agent during the melt processing to produce the foamed product. It has extensive use and in fact dominates the packaging industry. Because of its low cost, ease of molding and availability in FDA approved grades, polystyrene finds its way into many household and kitchen products.

5.2.5 Acrylics

Commercial acrylic is mainly polymethyl methacrylate (PMMA). It is an amorphous polymer and has excellent clarity. Because of its high resistance to UV light and good weathering property, it is widely used to make lighted sign boards and outdoor lighting articles. It is a good substitute for glass, except for its low scratch resistance. The disadvantage of acrylics is their poor resistance to organic solvents and low stress cracking resistance.

5.2.6 Polycarbonates (PC)

PCs are almost considered engineering materials because of their high impact strength. Their high transparency and high impact strength make them an ideal choice for automotive exterior lighting parts, helmet visors, and safety goggles. Processing of PC is fairly easy, although the presence of moisture can cause a lot of problems in the molding of PC. Cosmetic issues, such as splay, are common. A variety of PCs are available on the market and care must be taken to study the data sheet before molding a new grade of PC. The processing may be so widely different that the processing temperatures of two PCs may not even overlap. PCs have low stress cracking resistance and their chemical resistance is fair.

5.2.7 Polyesters

Polyethylene terephthalate (PET) and polybutylene terephthalate (PBT) are the most common injection molded polyesters. They exhibit excellent mechanical properties and are easy to process. By volume, PET is more commonly used in blow molding to produce containers and bottles. Because of their high chemical resistance to automotive fluids, PBT is used in under-the-hood applications. PBT is prone to be attacked by certain solvents.

5.2.8 Polyvinyl Chloride (PVC)

PVC is one of the least stable polymers, but is processed successfully with the appropriate additives and stabilizers. It is probably the material that has had the worst reputation over the years because of safety concerns. However, PVC is one of the most versatile polymers. It can be easily modified to achieve a wide range of flexibility. PVC by itself is very rigid and is marketed under the name rigid PVC or UPVC. PVC has excellent chemical and weathering resistance and is therefore widely used in pool and garden equipment. PVC has excellent electrical properties and is also used as cable insulators. The addition of plasticizers softens rigid PVC. With the addition of a large amount of plasticizer PVC can be processed into films by non-melt processing techniques. To make PVC more chemical resistant, it is chemically modified to incorporate chlorine in the molecule. This PVC is called chlorinated PVC. Heat resistance and weatherability decrease with the addition of the chlorine. Melt processing of PVC requires extreme care and caution. Decomposition of PVC produces hydrochloric acid which is an irritant to humans. Therefore, the residence time in the molding barrel must be controlled and limited. Once the production run is over, all the PVC in the barrel must be thoroughly purged. All related equipment must also be cleaned to avoid any of the PVC from getting to the barrel with the next material. This is especially true if the next material to be molded is an acetal. Acetal and PVC are known to be an explosive combination in their melt form and therefore any chance of them being mixed must be avoided.

5.2.9 Polyoxymethylene (POM or Acetal)

Although acetal is the commonly used term (sometimes polyformaldehyde is also used), this material is actually polyoxymethylene. It is a crystalline polymer and is considered an engineering plastic. It has excellent stiffness, fatigue endurance, resistance to creep, and low coefficient of friction. POM also exhibits excellent chemical resistance below 70 °C to all organic solvents. Above this temperature, some phenolic materials can react with the polymer. Resistance to inorganic polymers is low. Gears, bearings, conveyor parts, or any moving parts in assemblies are good application candidates for acetals because of their unique properties. The cost of acetal is fairly high. The density of the unfilled polymer is 1.42 g/cm^3, which is very high compared to olefins or nylons. This makes it less attractive when it competes with other materials because the part will weigh more and therefore cost more, if molded in acetal. Melt processing of acetals is fairly easy. Overheating of processing barrels and large residence times must be avoided. Degradation produces formaldehyde gas which is an irritant. Acetals are non-hygroscopic and therefore do not require drying. Surface drying is helpful in humid environments. A typical problem occurring when molding acetal is that the parts show unusual shrinkage values, far different from the published values. The main reason is usually due to the gate freezing off before the part is packed out with the required amount of plastic. Enlarging the gate size helps.

5.2.10 Fluoropolymers

Fluoropolymers are polymers containing fluorine. Polytetrafluoroethylene (PTFE), also called Teflon, is the most commonly used fluoropolymer. However, because of its high melt viscosity it is almost impossible to be injection molded. It has excellent mechanical properties over a wide range of temperatures. It does not dissolve in any acids, alkalis, or organic solvents. It also has a low dielectric constant. Because PTFE cannot be melt processed, the sintering technique is used to produce PTFE components, such as pump valves. PTFE tape commonly used before installing a water pipe on to a mold is produced by the process of skiving. Polyvinylidene fluoride (PVDF) is a fluoropolymer that can be injection molded. PVDF also offers good chemical resistance although it is not as high as that of PTFE. PVDF has been successfully used in chemical and electrical equipment and even in the manufacture of internal battery components. Perfluoroalkoxy copolymer (PFA), fluorinated ethylene propylene (FEP), and ethylene-tetrafluoroethylene (ETFE) are other fluoropolymers that offer similar properties as PTFE, but they are modified to be injection moldable. Molds that need to be designed for molding fluoropolymers need to have wide runners and gates to reduce the pressure drop through the mold. Special screws with generous section dimensions, large feed sections, and smaller metering sections can be helpful in the processing of these polymer.

5.3 Additives

Today, all polymers that are supplied by the manufacturer have additives blended into them. Additives can change the properties of the polymers substantially. PVC is the best example of this. As mentioned earlier, PVC by itself is rigid but with the addition of a plasticizer it can become extremely flexible. All additives must be compatible with the base polymer and must not bleed out or plate out under extended service conditions. Some PVC films get tacky with increase in temperature. This is the bleeding out of some of the plasticizers contained in it. They must also be able to withstand the processing conditions such as melt temperature and shear rates. These additives can be solids, liquids, gases, or other polymers. Some of the important additives based on their use are described in the following.

5.3.1 Fillers

Fillers are added to polymers either to improve the properties or to reduce the cost. Glass fiber is an example of a filler used to increase the stiffness or creep resistance of the base plastic, it is also called a reinforcing filler. Other examples of reinforcing fillers are wood flour, carbon fibers, and nylon fibers used to increase the impact strength and rigidity of the base plastic. Glass beads are usually added to plastics to reduce the shrinkage or sometimes (depending on the base polymer density and the glass bead density) the weight of the polymer. These are called inert fillers. Talc and calcium carbonate are other examples of inert fillers. In theory, any material that is compatible and can be blended into the polymer matrix can be used as a filler. Over the years, experiments have been conducted with a large number of fillers, especially when a particular material was easily available in the local geographic area. For example, fibers from the coconut tree or from the coconut have been successfully used in some applications in some South Asian countries. In some cases, coupling agents are used to form a bond between the filler and the polymer since the two are not always compatible, typically with inorganic fillers.

5.3.2 Plasticizers

Plasticizers are added to provide flexibility and softness to a material. Because of their compatibility, they get in between the polymer molecule, increasing the intermolecular distance. They also help in reducing the melt viscosity for improved processing properties. PVC is the biggest consumer of plasticizer by volume. Plasticizers are usually low molecular weight compounds, usually oligomers. If the plasticizers are not compatible, they tend to migrate to the surface of the polymer. Phthalates such as di(isooctyl) phthalate (DIOP) are commonly used as plasticizers in PVC. Concerns regarding safety have led researchers to more natural products and in recent years epoxidized soybean oil (ESO) has gained popularity.

5.3.3 Flame Retardants

For most products that are in direct contact or in close proximity with humans, such as furniture, toys, televisions, and computers, flame retardants are added to prevent the spread of fire. Although the plastics used in these products will burn in the presence of a fire, they must not continue to do so when the source of the fire is removed. Most consumer materials, such as olefins and polystyrenes, will continue to burn even if the source is extinguished. Flame retardants are added to plastics to inhibit this continued burning. This is achieved by one of several mechanisms, such as forming an incombustible gas and cutting of the supply of oxygen required for the burning. Halogen-containing compounds, such as chlorinated paraffins, and phosphates, such as trixylyl phosphate, are used as flame retardants. Testing is done using various methods and the rate of burn, time of burn once the flame is removed, and the amount of burn of the sample are all recorded as vital data.

5.3.4 Anti-Aging Additives, UV Stabilizers

Over time, polymers tend to lose some of their properties because of a variety of changes in the polymer: The molecular weight can drop as a result of chain scission or certain additives can degrade or bleed out. Some of these can cause chain reactions because the reaction products can help in the propagation of the reaction. Antioxidants and UV stabilizers are commonly added to plastics as anti-aging additives. The antioxidants react with any free radicals that are formed and prevent the propagation of the chain scission. UV light can cause a change in the physical properties of the polymer, e. g., causing it to become hard leading to surface cracking. They may also discolor. Automotive head lamp covers tend to become yellow over time because of their extended exposure to the UV light present in sunlight. UV stabilizers absorb the UV light and prevent it from affecting the plastic. Certain amines are examples of UV stabilizers. All products that are exposed to sunlight must have a UV stabilizer incorporated.

5.3.5 Nucleating Agents

Nucleating agents are used in crystalline polymers to help speed up the nucleation process and lower molding cycle times. When the melt enters the mold, it is completely amorphous and as it starts to cool down, the crystallization process starts. Mold temperatures are maintained high enough to supply the energy for crystallization. Crystallization provides the required properties to the end product and therefore it must be completed before the plastic temperature falls below the glass transition temperature (T_g). Below the T_g, no meaningful crystallization can take place. Nucleating agents help to increase the onset and rate of nucleation. They are commonly incorporated in olefins, polyesters, and nylons that form the majority of the injection molded crystalline polymers. Talc is a good example of a nucleating agent for polypropylenes.

5.3.6 Lubricants

Lubricants are added to reduce the coefficient of friction in the final products. In many assemblies, moving parts slide against each other (cams) or drive another part by an applied mechanical force (gears). In such cases, a low coefficient of friction is desired for two reasons:

- the energy required to move the parts is reduced and
- a lower coefficient of friction translates to less heat buildup in the part. Heat buildup can cause thermal expansion of the component changing its dimensions and leading to the failure of the assembly.

Graphite and PTFE are added to plastics in small quantities to act as lubricants.

5.3.7 Processing Aids

Some texts classify lubricants and processing aids in one category. Processing aids act as melt viscosity reducers that help in the melt processing of the plastic. They are also called flow promoters. Amine waxes are examples of these. Processing aids do not alter the final property of the plastic and their primary function is to ease the processing.

5.3.8 Colorants

Colorants are blended into the base polymer to impart the destined color. Compatibility is no doubt important, but in injection molding, where color pellets are mixed with the plastic, the carrier resin for the colorant also needs to be compatible. Colors can also be added as liquids directly into the barrel of the injection molding machine, although this can get messy. The advantage here is that the control of the amount of liquid can be very accurate. Mixing color powders with the resin must be done immediately prior to processing because the powder tends to separate during transport. The powder also tends to stick to the walls of the mixing equipment and hoppers.

5.3.9 Blowing Agents

Blowing agents are used to mold plastic parts with a cellular structure. In thick parts this helps reduce the overall weight of the part and it also improves the structural strength of the part. Defects related to thick sections, such as sink marks, are eliminated. The blowing agent mixes with the plastic and creates internal pressure to push the plastic towards the walls of the mold. The molded parts are therefore dimensionally more consistent as compared to conventional parts. The disadvantage of using a blowing agent is the additional cost and effort of mixing. Parts molded with a blowing agent do not have good surface finishes. Large hoppers or containers are examples of parts molded using blowing agents. When the process is used to improve the structural properties, it is also called structural foam molding. Styrofoam cups are molded using a blowing agent. Blowing agents can be chemicals that decompose in the injection barrel to form a gas and help foam the plastic (e. g., sodium bicarbonate) or they can also be gases, such as nitrogen and air.

5.3.10 Other Polymers

Technically, when polymer blends are produced, one polymer is added to another polymer. Therefore these polymers may also be considered as additives. For example, in a PC/ABS

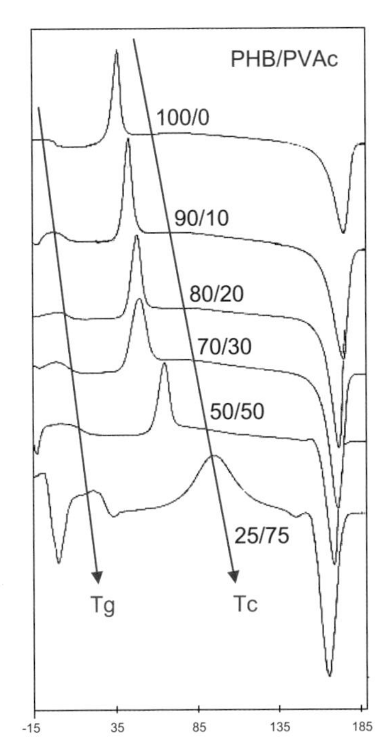

Figure 5.3 Thermal properties of blends of PHB/PVAc

blend, PC can be considered an additive to ABS or vice versa. Thermal properties of a blend of polyhydroxybutarate and polyvinyl acetate are shown in Fig. 5.3.

5.4 Closing Remarks

In Chapters 2 and 3, the morphology and the rheology of the polymers were discussed. Additives can have a profound influence on both these properties. A polymer that is naturally crystalline can be made completely amorphous by incorporating additives. Figure 5.3 shows the thermal properties of a PHB/PVAc blend. The PVAc was blended in various proportions into the PHB starting with 10 %. As the amount of PVAc increased, the glass transition temperature and the crystallization temperature of the PHB phase increased. Polymer properties can be tailored by blending techniques and by the addition of additives. However, there may not always be a win-win situation for all the properties. Some of them may have to be compromised.

Reference

1. Alfrey, Turner, and Gurnee, E.F., Organic Polymers, Prentice-Hall, 1967, p 51.

Closing Remarks

References

6 Injection Molding

6.1 The History of Injection Molding

The concept of injection molding for plastics was adopted from the metal die cast industry, which began to develop in the early to mid 1800s. Plastics were not known for their useful properties at this time. All polymers were naturally occurring materials and did not have much industrial value. In 1869, John Hyatt developed a concept to make billiard balls from cellulose nitrate, also known as celluloid. His machine consisted of a cylinder heated by steam to melt the celluloid and a hydraulic plunger to inject the plastic into the mold. Developments continued and several inventors produced several designs of molding machines and their components. Over the years, molding machines advanced from their manual operations to automatic operations with electrical controls. Almost seventy years after the first injection molding machine was developed, a major milestone was achieved. In 1948, for the first time, the two-stage screw was introduced in the injection molding industry. Until then melt homogeneity was achieved with the help of torpedoes in the injection barrel. The use of the two-stage screw provided the required homogeneity and allowed better control of the plastic shot size injected into the mold. As a result of the improved melt efficiency and better melt homogeneity provided by the two-stage screw, a larger volume of plastic could now be melted, leading to the possibility of molding larger parts. As molding machine control systems improved with the advent of advanced electronics engineering, the molding machines became more sophisticated, providing better control of the injection molding process. Multi-shot molding machines, better screw designs, clamp position versatility, and non-hydraulic all-electric machines were some of the many improved features and products being introduced to the molding industry. Today's machines have become highly complex and are capable of molding almost anything that a product designer would desire, from micro-molded parts used in the watch industry to large parts used in the automotive industry and everything in between. The latest generation of machines can be connected to any computer in the world via the Internet and monitored remotely. This helps with quick debugging and problem solving of machine and molding problems without requiring an engineer on site.

6.2 Injection Molding Machines and Their Classifications

A conventional injection molding machine is shown in Fig. 6.1 a. Plastic pellets are fed into the barrel of the injection molding machine where shear heat generated by the rotating screw and external heat provided by electric heaters around the barrel melt the plastic, making it

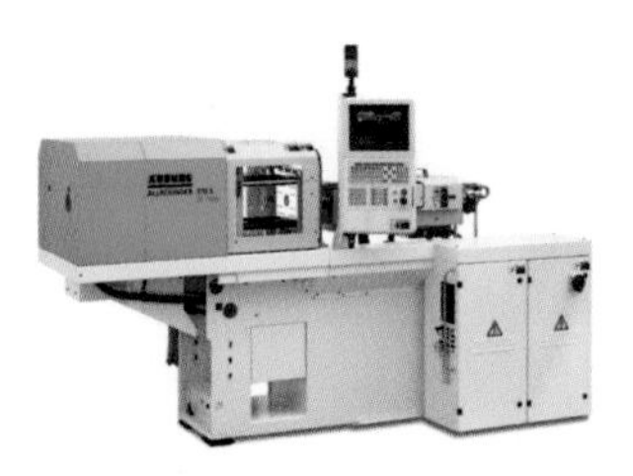

(a) Horizontal Clamp – Horizontal Injection

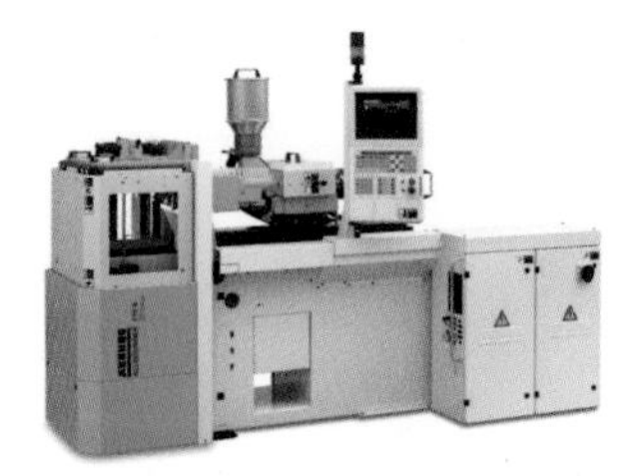

(b) Vertical Clamp – Horizontal Injection

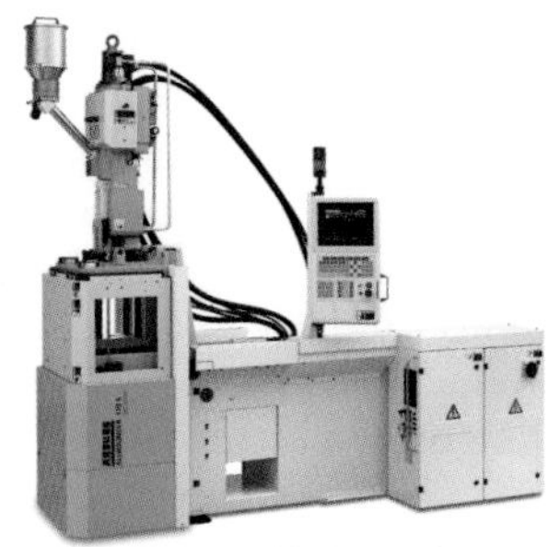

(c) Vertical Clamp – Horizontal Injection

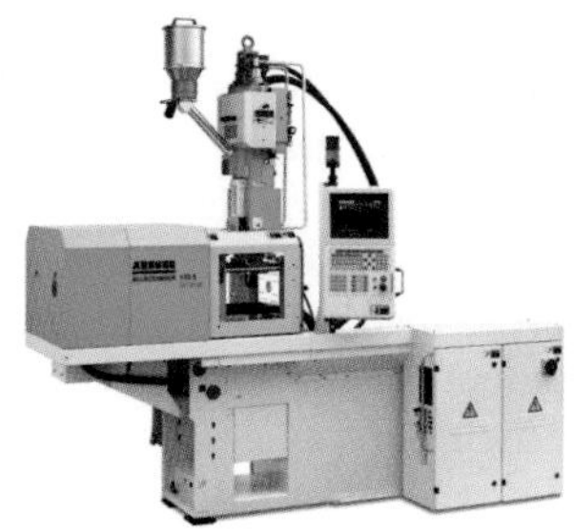

(d) Horizontal Clamp – Vertical Injection

Figure 6.1 Machine classifications based on clamp and nozzle positions (Courtesy: Arburg Inc.)

ready to be processed. As the screw rotates, it augers, the required amount of plastic for the shot to the front of the barrel. This plastic is then injected into the mold by the forward movement of the screw. The mold has a coolant flowing through it that helps maintain its temperature. Once the part is cooled below its ejection temperature, it is ready to be ejected out of the mold. Although the process of injection molding seems fairly simple, there are multiple speeds, pressures, times, and temperatures that must be controlled in order to produce quality products. It is the optimization of these processing parameters necessary to run an efficient and successful molding operation that Scientific Processing is all about.

Injection molding machines are classified in a number of ways. Based on the movement of their clamp, they are classified as horizontal clamp or vertical clamp molding machines (see Fig. 6.1 a–d). Horizontal clamp machines are suitable for most applications where the parts are ejected out of the mold and fall onto a conveyor or into a box due to gravity. Molded parts may then go through a secondary operation before being packed and sent to the customer. Horizontal clamp machines are the most versatile machine and are most common. Vertical clamp machines are suitable for producing insert molded parts, where a horizontal parting line is an advantage to seating inserts into the mold. It is the vertical clamp orientation of a vertical molding machine that allows the mold parting to remain parallel with the floor. This arrangement allows gravity to hold inserts or other components to be over-molded into position in the lower half of the mold. Once the inserts are positioned, the top half of the mold can be clamped into position over the inserts that have been placed into the bottom half. Vertical molding machines commonly have a rotary or a shuttle mechanism that moves the bottom half of the mold from underneath the clamping mechanism to an area easily accessible by either an operator or a robot for unloading the parts and loading the inserts. In most cases there are two or more moveable mold bases. While inserts are being loaded into one of the exposed mold bases, another mold base is clamped shut and being injected with plastic. This parallel activity saves time and increases output of the insert molding process. It also helps in reducing residence time of the plastic in the barrel.

Injection molding machines may also be classified based on the direction of plastic injection. Horizontal injection machines are most common. Here, the injection unit is mounted parallel to the ground and plastic is injected through a sprue bushing usually mounted in the center of the mold. Plastic flows into the mold via the sprue bushing, flowing through the center of the mold half, perpendicular to the parting line of the mold (i.e., the split line where both mold halves separate). The runner system delivers plastic to the cavities and is cut into one or both sides of the mold. The sprue intersects the path of the runner in the center of the mold, allowing plastic melt to be distributed to the cavities. Upon mold open, the molded parts and runners remain on the moving side of the mold half. In this scenario, the hardened plastic sprue is pulled out of the stationary mold half, leaving a long appendage of plastic that can be gripped by the jaws of a robotic sprue picker as part of the part ejection cycle. This is the standard horizontal molding arrangement. On the other hand, some molded components with part design and/or mold design restrictions, cannot use a sprue and runner configuration for delivery of plastic through the center of one mold half. Instead, the sprue is eliminated and the entire runner system has to be at the parting line. This is typically referred to as a parting line shot, since the plastic is injected into the mold where the mold halves meet, on the parting line. In such cases vertical injection of the plastic is required and therefore the injection barrel is rotated to a vertical position. There is a limitation on the size of the vertical injection unit because it is usually supported on tie bars. In some cases, a vertical injection unit will have its own independent support units, but these still have to be versatile in their horizontal movement to allow adjustment for the position of the parting line on the mold.

Multi-shot machines have two or more injection barrels and are capable of molding parts that are comprised of two different materials. A toothbrush is a common example of a part

Figure 6.2 Two shot (material) machine (Courtesy: Arburg Inc.)

produced by a two shot process. With the first shot, the plastic substrate is molded. The mold is then opened and the cavities with the first shot of plastic are indexed to align with the second injection unit. When the mold closes, the second shot is injected over the now hardened first shot or substrate, creating a two shot component that is ejected from the mold. Typically, the second shot is a soft elastomers-type material typical for the soft handle molded over the toothbrush substrate. A two-shot machine is shown in Fig. 6.2.

6.3 Machine Specifications

Molding machines are most commonly specified by their tonnage and shot size. When selecting the molding machine, these two are the primary specifications that are considered. Once these requirements are satisfied, other specifications are considered.

6.3.1 Clamp Force (Tonnage)

The plastic is injected into the mold under high pressure. This pressure generates a force on the mold face that can very easily open the mold during injection. To prevent the mold from opening, the clamp is kept locked down by a counterforce. The maximum available force to keep the mold closed is called the clamp force of the machine. It is usually specified in tons and is therefore also called the tonnage of the machine. Appendix B lists the different units that are used to specify machine tonnage.

6.3.2 Shot Size

Shot size is another important parameter to consider when selecting a molding machine. The shot size of a molding machine is determined by the maximum weight of general purpose polystyrene (GPPS) that can be molded with a single stroke of the injection molding screw. A shot size of 100 g would indicate that a GPPS part weighing no more than 100 g can be molded on the machine. It is becoming more common to use the volume of the cylinder that comprises the barrel of the molding machine rather than the GPPS part weight because the latter can be misleading. The density of GPPS is 1.06 g/cm^3 and therefore, for the given volume of the cylinder/barrel of the molding machine, it can hold more or less weight of another material, depending on its density. To make matters simple, consider the shot size of the machine to be 106 g, which equates to 100 cm^3 of volume. Therefore, it can hold 106 g of GPPS. If this GPPS was replaced with low density polyethylene (LDPE) with a density of 0.91 g/cm^3, the maximum shot size now is 91 g of LDPE. For a 30 % glass filled nylon (density of 1.33 g/cm^3), the shot size is 133 g. It is therefore now becoming common to specify the shot size in terms of volume rather than weight. The setting of the shot size on the machine is also taking this approach because it helps when moving the mold from one machine to another machine of a different screw diameter. It is easier to match the shot volume than calculate the new shot size in linear dimensions.

6.3.3 Screw Diameter and L/D Ratio

As the name suggests, the screw diameter is the diameter of the screw, either in millimeters or in inches. The barrel diameter is only slightly bigger than the screw. For example, for a 50 mm screw, typical barrel clearance is 0.1 mm (4 thousands of an inch). Over time, screw wear can cause material leaking over the screw flights, causing inconsistency from shot to shot.

The ratio of the length of the screw to the diameter of the screw is called the L/D ratio. The higher the L/D ratio, the better is the melt homogeneity.

6.3.4 Plasticating Capacity

The plastic has to be melted and heated to its processing temperature. The melting of the plastic is accomplished by the heater bands around the barrel and the shear heat generated by the rotating screw. The screw is also transporting the plastic to the front of the barrel, ready to be injected into the mold. Barrel temperature, exposure time, and screw speeds are all important to get a good homogeneous melt. If the material is moved too fast, the plastic pellets may not have enough time to melt and the melt can contain plastic pellets that are not or only partially melted. During the purging process, it is common to see un-melted pellets of polyethylene coming out of the nozzle tip if high screw speeds were used. Plasticating capacity is the maximum weight of GPPS that can be raised to the molding temperature and metered in front of the screw. It is usually expressed in kilograms or pounds per hour.

6.3.5 Maximum Plastic Pressure

Thin-walled parts require more pressure to fill than do thick-walled parts. Therefore, depending on the application, the maximum available pressure needs to be specified. For hydraulic machines, the maximum available plastic pressure depends on the maximum available hydraulic pressure and the intensification ratio of the screw. Plastic pressure is specified in pounds per square inch or bar or other pressure units.

There are other parameters that are specified on machine data sheet. Discussions of these are beyond the scope of this book. Suggested books on this topic are mentioned at the end of this chapter.

6.4 The Injection Molding Screw

The injection molding screw and barrel assembly is responsible to help deliver the right quality of melt to the mold. Electric heater bands are installed around the barrel and supply radiant heat energy to melt the plastic. However, it is the screw that plays the vital role in the process of achieving a homogeneous melt. The screw provides shear heat to assist in the melting process along with the required mixing and homogenizing of the melt. It also

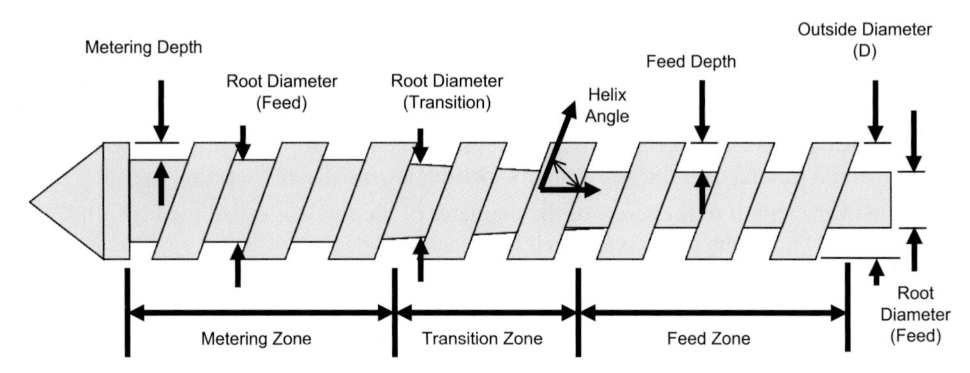

Figure 6.3 The general purpose screw

helps in accurately measuring the volume of the shot to be injected into the mold. Several designs of screws have been developed based on different materials and their needs. The most common screw is called a general purpose screw or the GP screw. The design of a GP screw is shown in Fig. 6.3.

The GP screw has three main sections (or zones), each of which serves a special purpose. They are described in the following together with some other terms. The screw construction can be described as a shank with flights that are wrapped around it. In most cases, there is only one ribbon that creates the flights.

Outside Diameter: This is the diameter of an imaginary cylinder that is created by joining the outside area of the flights. The outside diameter is constant and is slightly smaller than the internal diameter of the barrel.

Root Diameter: This is the diameter of the shank. The root diameter changes from the back to the front of the screw, depending on the section of the screw.

Channel Depth: The difference between the outside diameter and the root diameter is the feed depth. Since the root diameter changes, the feed depth also changes from the back to the front of the screw.

Feed Zone: This is the section of the screw that picks up the material from the feed opening (base of the hopper) and begins to soften the material as it is being conveyed. The root diameter is the smallest here and is constant. Since the root diameter is constant, the channel depths are also constant; they are also called the feed depth. In the feed zone, the material is picked up and is softened as it is conveyed by the rotation of the screw. The material must never be completely molten in this section because that would prohibit the picking up of additional material. A term commonly used to describe this phenomenon is *screw slipping*, where the melt rotates with the screw and prohibits the screw from moving back to pick up more material and build the next injection shot.

Transition Zone (Compression Zone): In this section, the root diameter increases gradually, resulting in the decrease of the channel depth. At the start of this section, the root diameter is the same as the root diameter of the feed section, where it gradually increases until the section ends. This causes the feed depth to steadily decrease. As the screw rotates, the softened pellets begin to get compressed and the air and any other volatiles are forced out from between them because the feed depth is decreasing and the plastic is being conveyed. With the help of heat from the external heater bands and the shear from the rotation of the screw, the plastic begins to melt. As the feed depth reduces because of dispersive and distributive mixing, the plastic ends up as a homogeneous melt by the time it reaches the end of the transition zone. Distributive mixing occurs when the melt streams distribute and reconvene. On the other hand, dispersive mixing is similar to a smearing action. There is a combination of distributive and dispersive mixing taking place in the barrel.

Metering Zone: The metering zone is the last zone and is the closest to the nozzle of the machine. The depth of the channels in this section is minimal compared to the other two sections. The root diameter stays constant and therefore the channel depth is also constant. Since the shot is built by moving the screw back until it reaches a set linear position (shot size), the metering depth must be as minimum as possible to reduce the variation in the amount of melt for each consecutive shot. With a larger metering depth, the amount of material that is fed in front of the screw can vary, leading to inconsistencies. However, as the depth reduces, the shear increases and therefore the risk of material degradation is also increasing, especially for shear sensitive materials such as PVC. A compromise must be found and special screw designs are therefore necessary for certain types of materials.

Figure 6.4 shows the melting progression of the plastic as it travels through each of these sections. In the feed zone, the pellets have softened and begin to adhere to each other. When they travel to the transition zone, there is a combination of melted and un-melted plastic. There is still evidence of plastic pellets that have been compressed together. The metering zone shows a ribbon of completely molten plastic. In a GP screw, the length of the metering and transition zones are the same and the feed zone is usually twice the length of any of one these sections. In custom designed screws, these lengths can be altered. Longer feed zones increase the throughputs, longer transitions decrease shear, and a longer metering section will output a more homogeneous melt but will create more shear.

Compression Ratio: This is the ratio of the feed section channel depth to the metering section channel depth. It defines the amount of compression to which the material has been

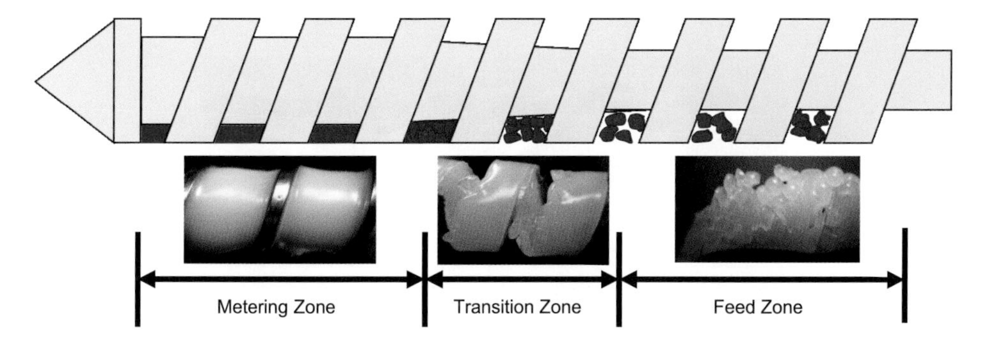

Figure 6.4 Melting progression of the plastic as it travels through the sections of the screw

subjected. The higher the compression ratio, the better is the melt homogeneity, but also the higher is the shear. The depth of the channels also contributes to the amount of shear heat, melt homogeneity, and the throughput.

Typical compression ratios are mentioned below:

- Low compression ratio: 1.5:1 to 2.5:1 used for shear sensitive materials such as PVC
- Medium compression ratio: 2.5:1 to 3.0:1 used in general purpose materials
- High compression ratios: 3.0:1 to 5.0:1 used for crystalline materials such as nylons

Helix Angle: This is the angle of the flight in relation with the plane perpendicular to the screw axis.

L/D Ratio: The L/D ratio is the working length of the screw flight to the outside diameter of the screw. Most injection molding screws have L/D ratios of 20:1. Greater L/D ratios allow more exposure of the plastic to heat and shear, improving the melt homogeneity and therefore increasing throughput at the desired processing temperature.

6.5 Screw Designs

The screw described in Fig. 6.3 was a general purpose design that can be used for most materials and in most situations. Screws have been designed for specific requirements, such as for shear sensitive materials or to improve the throughput to reduce cycle times. Screw designs also depend on the degree of crystallinity, the viscosity, and the additives present in the plastic. In injection molding, mixing screw designs and barrier screw designs are most common. Mixing screws, as the name suggests, help to mix the additives such as colorants and also help to improve melt homogeneity. There are certain sections incorporated into the screw that create the mixing effect. Barrier screw designs have two screw channels in the transition section of the screw separated by a barrier flight. The un-melted plastic stays in the first channel until it is completely molten and then moves on to the second channel. This ensures that the plastic is completely molten before it reaches the metering section. Examples of screw designs are shown in Fig. 6.5.

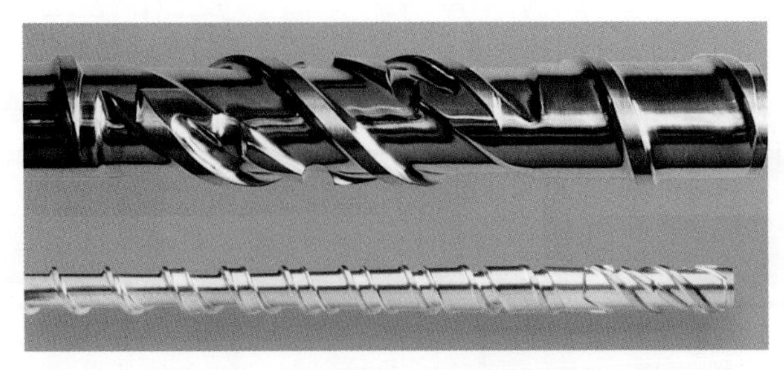

Figure 6.5 A mixing screw and a barrier screw (Courtesy: Westland Corporation)

6.6 The Check Ring Assembly

The check ring assembly is essentially a non-return valve. The most popular design is shown in Fig. 6.6. During the shot build-up process, the screw is rotating to pick up the material. During this time, the check ring is in the forward position and allows the passage of the material to the front of the screw. During injection, the check ring seats itself on the body and stops the plastic from being pushed back over the flights, acting as a one-way non-return valve. Check rings wear over time and the plastic will begin to leak into the flights. This causes inconsistency in the shot and therefore shot-to-shot variations. Check rings must be checked periodically and changed at the slightest sign of leakage. Various designs of check rings are available on the market for specific materials and applications.

Other types of check rings include ball check valves that are suitable for unfilled, non-shear sensitive materials such as polyolefins. Smear valves are used to process highly viscous materials such as rigid PVC. There is no shut-off mechanism in a smear valve. The high viscosity helps to prevent the back flow of the plastic during injection.

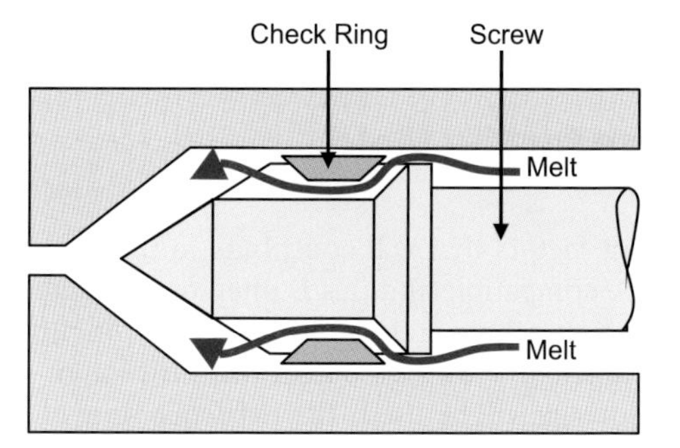

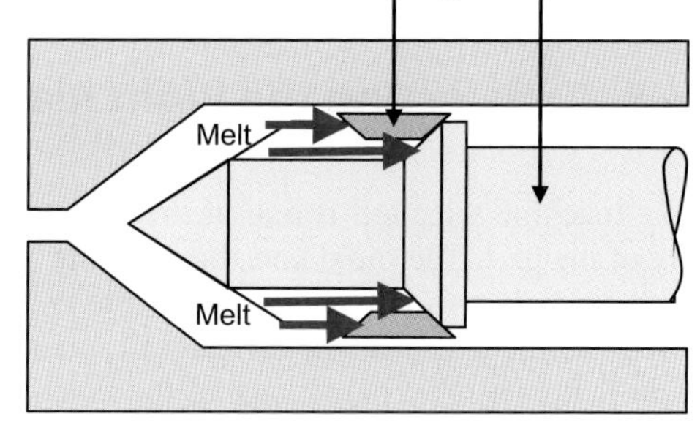

Figure 6.6 Working of a check ring

6.7 Intensification Ratio (IR)

To drive a nail into a wall, the force is applied to the head of the nail. The cross section of the area at the head of the nail is comparatively larger than the point. Therefore, the applied force of the hammer at the head of the nail is intensified at its point and this is what pushes the nail into the wall. In injection molding, there is a hydraulic ram at the back on the molding screw where hydraulic pressure is applied. This force is intensified at the screw tip and converted to plastic pressure. The ratio of the cross sectional area of the ram to the cross sectional area of the screw is called the intensification ratio (IR), see Fig. 6.7. Plastic pressure at the nozzle is equal to the product of the applied hydraulic pressure and the IR. Intensification ratios range from about 6:1 to as high as 23:1. Higher IRs are necessary to mold thin parts with long flow lengths. In electric machines, there is no hydraulic ram and the pressure applied to the back of the screw is what the plastic experiences so that the IR for an electric machine is equal to 1.

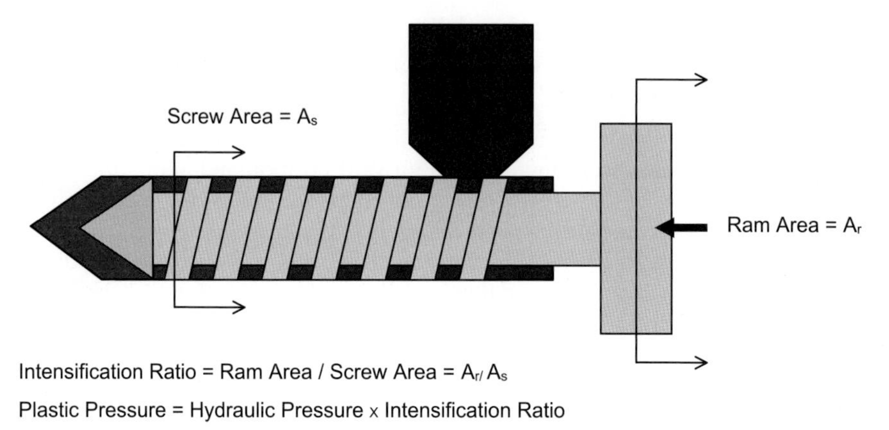

Figure 6.7 Intensification ratio

6.8 Selecting the Right Machine for the Mold

The machine selection is one of the five important factors that will contribute to the quality of the part. The mold and the machine must be compatible and this is often overlooked. Usually, only two factors are taken into consideration: whether the mold physically fits in the machine and whether the clamp tonnage is sufficient. However, one of the most important factors is what percentage of the machine shot size will be used. The residence time of the material in the barrel and the lower limiting size of the mold are other factors. Both are described in the following.

6.8.1 Physical Size of the Mold

The mold size is defined by three variables.

Mold Stack Height (H): This is the distance between the sides of the mold in the direction of the mold open and close when the mold is fully shut.

Mold Width (W): This is the distance between the vertical sides of the mold, looking in the direction of injection of the plastic. This applies to molds mounted in horizontal machines, but the definitions are extended to other molds also.

Mold Length (L): This is the distance between the top and bottom side of the mold, looking in the direction of injection of the plastic. This again refers to horizontal molds.

Naturally, the mold must fit in the machine such that at least two sides can be bolted to the platen of the machine. The mold can hang off the platens on the other two sides, but the molding area must not be outside the platen. The molding area must always be supported by the platen, see also Fig. 6.8. If the cavity is not supported, the injection pressure can easily deflect the plates and cause flash in the part. Injection pressures can be very high, applying tremendous force on the mold base, and over time can damage the mold components, if the mold is not properly supported by the mold platens. On the other hand, the mold must not be considerably smaller than the platen. It must cover at least 70 to 75 % of the area between the tie bars. This is especially true for toggle machines, where the clamping force is applied on the outside and not in the center of the platens. There is a possibility of platen deflection, if the mold is too small causing platen damage over time. The toggle system also provides reduced support in the center of the mold where the main injection pressure is applied. Even with adequate support pillars in the mold, there still could be deflection because of the lack of support, causing part defects.

In regards to the mold height, every machine has a minimum and a maximum mold height that it can accept. The moving platen closes the mold and applies the set tonnage to the mold. Because of the limit on the travel distance during closing, the mold height needs to be greater than this limit. If the mold is less than the minimum mold height, the platen can never let the two halves of the mold touch and apply the tonnage. Therefore minimum mold height is important. On the other hand, the mold must be smaller than the maximum mold height in order to fit in the machine.

The required mold open stroke will depend on the part, usually the part dimension in the direction of ejection. The open stroke should be such that the mold halves are far enough apart when fully opened so that the part falls out of the mold after ejection. The mold open stroke must be greater than the longest dimension of the part in the direction of ejection. For example, on a rectangular part as shown in Fig. 6.9, it should be the diagonal of the part. Even with this distance, there is still a danger of damaging the part because it may still hit the side of the mold as it falls off the ejectors. For this reason, the mold opening stroke must be set as wide as possible to avoid damage, but not such that cycle time is lost due to unnecessary movement of the mold.

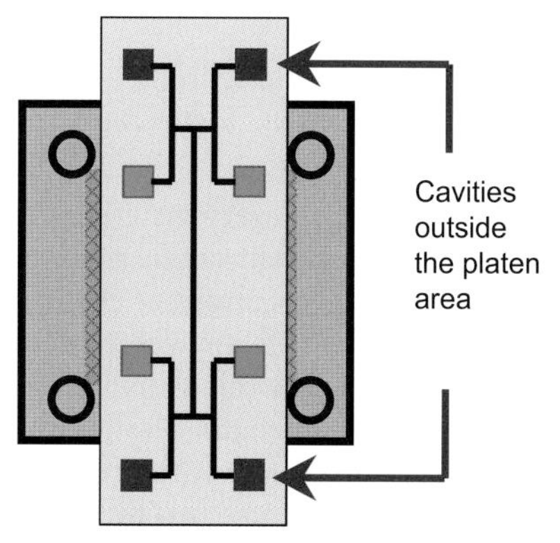

Figure 6.8 Cavities outside the molding area

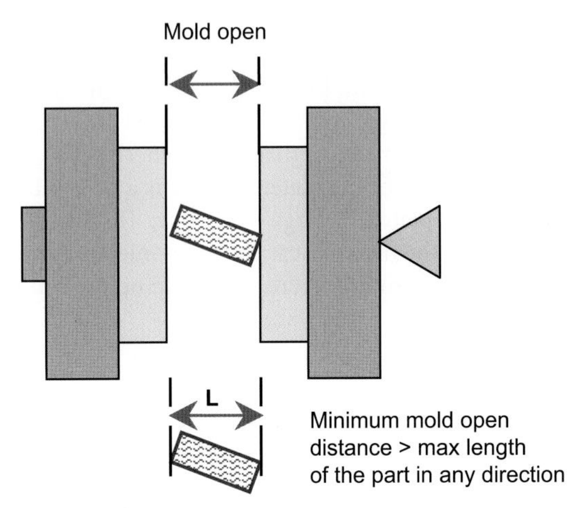

Figure 6.9 Minimum mold open distance

6.8.2 Tonnage of the Machine

The injection pressure of the plastic applies an outward force on the mold cavities, which works to separate the mold halves. This force must be counter balanced by the machine in order to keep the mold halves closed. If the plastic pressure is higher than the clamp force applied to keep the mold closed, the mold will open and the plastic will escape from the mold at the parting line where the mold splits open, causing part defects, typically flash. The force that keeps the mold closed is called the clamp tonnage of the machine. The applied clamp force is measured in tons. The rule of thumb for calculating the tonnage required for the part is given in Eq. 6.1.

$$\text{Required tonnage} = (\text{Projected area of the part} \times \text{Number of cavities} + \text{Projected area of the runner}) \times (\text{Tons / in}^2 \text{ required for the resin}) \quad (6.1)$$

Projected area is the 2-D area of the part looking in the direction of injection or the plane view of the part in the direction of injection. Figure 6.10 shows the shaded area as the projected area of a part.

Depending on the plastic material properties, every material exerts a certain amount of force during the mold fill and then requires a certain amount of pressure to pack the part out. Typically, crystalline materials require 3.5 to 4.5 tons of clamp force per square inch of projected area, while amorphous materials require anywhere between 2.5 to 4.0 tons of clamp tonnage per square inch of projected area. The calculation is only a rule of thumb and other factors, such as wall thickness, plastic flow length, plastic flow direction, and presence of slides in the mold can have a significant effect on the required tonnage. In Fig. 6.10, the same part is shown to have injection points from two different directions. Using the above formula, the required tonnage when the plastic is injected from the side will be lower than when injected from the front. This does not mean that the part can be run on a lower tonnage press. The flow length would then play a role in the tonnage.

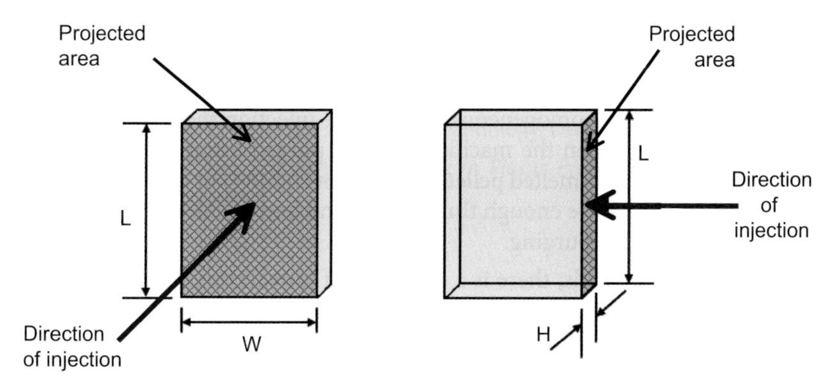

Figure 6.10 Part projected area in the direction of injection

Tonnage is only required when the mold is filled. High injection pressures in an incompletely filled mold will not exert any significant force on the clamping unit. It is only when the mold is full that any additional input of material will cause the force to be exerted on the clamp. Therefore, in thick parts it will be the pack and hold pressures that will dictate the required tonnage in the mold. On thin-walled parts, depending on the length of flow, there could be an exception and during the filling process, even though the part is not filled, a force may be applied on the clamping unit.

Tonnage calculation is very complex and not easy to predict. Computer simulation programs do an acceptable job in this calculation, but caution is warranted when applying the results.

6.8.3 Percentage Shot Size Used

For molding consistency, the percentage of shot size used is the most important factor and is often overlooked. The percentage shot size gives an idea of the amount of plastic injected into the mold with respect to the maximum amount of plastic the barrel can hold or the fraction of the shot size that is injected into the mold. The formula for this calculation is given in Eq. 6.2.

$$\text{\% Shot Size used} = ((((\text{Part weight} \times \text{Number of cavities}) + \text{Runner weight})) \times (1.06/(\text{Density of the plastic})))/(\text{Shot size of the machine})) \times 100 \tag{6.2}$$

The percentage shot size used must be always between 20 and 80% of the available shot size. A number below 20% results in inconsistency of the amount of plastic being injected into the mold. The screw pressure and velocity need some time to build and achieve the set pressures and speeds. If the shot is very small, this phase can be inconsistent because not enough time is given for pressure and velocity to reach required levels. Moreover, the plastic that has now built up the pressure is suddenly stopped and the momentum is unpredictable, leading to large variations in the fill. In general a smaller shot will result in larger variation. The shot size must therefore be larger than 20%. To ensure a more accurate fill, the minimum the minimum shot size should at least reach about 30–35%.

7 Scientific Processing and Scientific Molding

7.1 Introduction

Several parameters determine a successful molding process. There are various speeds, pressures, times, and temperatures to be considered. Scientific processing encompasses an understanding of the underlying scientific principles of each parameter and the application of these principles to achieve a robust process and consistency in part quality. Scientific processing covers the complete molding process, from the time the plastic enters the facility to when it leaves as a finished product. A robust process is one that can accept reasonable natural variations or a small purposeful change in an input but still delivers consistent output. The term consistency means molding parts with the least variation in the quality of the part. The quality of the part can mean its dimensions, appearance, part weight, or any other aspect that is important to the form, fit, or function of the part. The variation should be from special cause variations and not from any natural cause variations. Special cause variations are variations that are caused by an external factor. For example, if the chiller unit shuts down, the mold temperature will change causing a change in the quality of the part. Natural cause variations are inherent to the process. Their effect can be minimized but not eliminated. For example, if the plastic used to mold the parts has 30 % of glass fiber mixed in it, in every molded shot the amount of glass will not be exactly 30 %. It will be slightly more or less, for example, between 29.7 and 30.3 %. This variation cannot be eliminated, but the mixing process can be improved and the variation can be reduced.

Robustness and consistency should not be confused with parts being molded within the required specifications. Parts can be out of specifications but the process can be robust and the quality can be consistent. The goal of scientific processing is to achieve a robust process at each stage of the molding process the pellet is subjected to.

The term Scientific Molding was coined by a two pioneers in the field of injection molding, John Bozzelli and Rod Groleau. Their principles and procedures are widely used today and are industry standards. Scientific molding deals with the actual plastic that enters the mold during the molding operation at the molding machine. The term introduced here is Scientific Processing, which is defined as the complete activity the plastic is subjected to from the storage of the plastic as pellets to the shipping of the plastic as molded parts. Scientific processing is applying scientific principles to each of the steps involved in the conversion of the plastic to the final product, see Fig. 7.1. This chapter deals with the understanding and optimization of this complete process. The focus is on the understanding and the application of the theories to each of these steps and then optimizing them. Successful process development results in a process that is robust and one that molds parts with the required consistency.

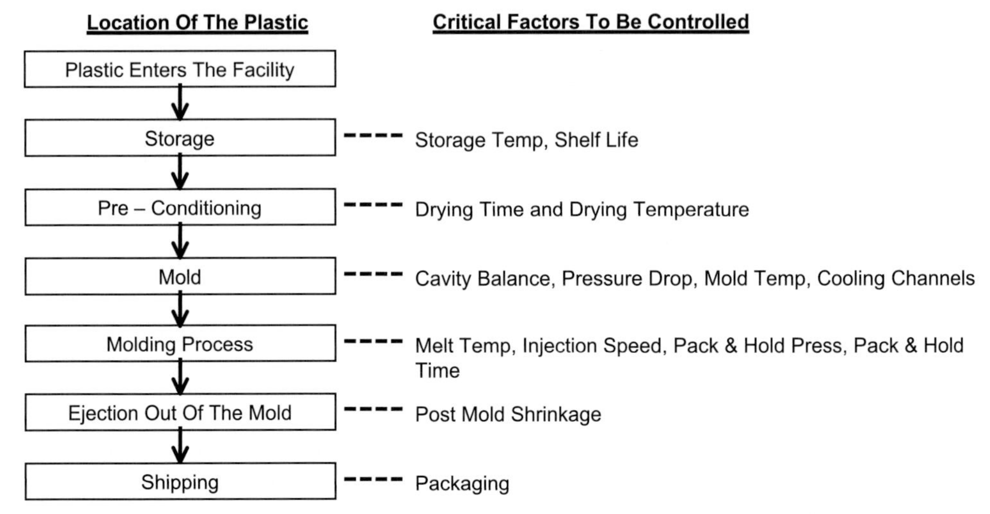

Figure 7.1 The journey of the pellet and the critical factors that need to be controlled

7.1.1 Process Robustness

A process is considered robust when changes to the inputs have minimum effects on the quality of the part. The changes here can be intentional or may be due to natural variations. Naturally, intentional changes must be within reason. In general, a process becomes more robust as larger input changes can be introduced without adversely affecting the resulting output part quality. For example, after a certain injection speed is reached, the viscosity of the plastic remains constant. The viscosity curve is in a robust area and variations in injection speed have little effect on the viscosity and therefore the amount of fill into the mold. At low injection speeds, a slight change in the injection speed causes a large change in the viscosity, resulting in shot-to-shot fill inconsistency. Therefore, this is not a robust area of the process and should be avoided. In addition, it must be understood that natural variations can never be eliminated. Taking these conditions into consideration will help ensure building a robust and consistent process.

7.1.2 Process Consistency

A process is considered consistent when it meets the following two requirements.

- All variations in the outputs of the process are a result of only natural cause variation.
- The standard deviation of the variation is at a minimum value.

For example, the cushion value is an output of the injection, pack, and hold phases. If the cushion value shows minimum variation, and a distribution curve of the cushion value over time is normal, then the process is consistent. In this case, the process under consideration would include only the injection, pack, and hold phases.

A robust process will always produce parts of consistent quality because there is little variation in the output. It also goes without saying that for the quality to be consistent, the process must be robust. For injection molding, whenever there is an inconsistency in part quality, the robustness of the process is usually suspect because the process is reflected in the part quality. In general, based on how robust the process is and on the required tolerance limits, we consider four possible resulting production process scenarios, as shown in Fig 7.2, which shows a representation of a run chart for a particular dimension.

Figure 7.2a shows a process that is not robust because of dimensional changes in the the part. The first four data points are closer to the upper specification limit, but the next data point drops down towards the lower specification limit. There are some parts being molded out of specifications. Figure 7.2b shows the same process, however with increased tolerances; this process is producing parts within specifications. In both these cases, a special cause variation seems to be contributing to the inconsistent part quality. An attempt must be made to eliminate this variation even though in the second case, the parts are within the required specifications. Figures 7.2c and 7.2d both represent a robust process, because the quality distribution is normal. In Fig. 7.2c, the tolerance limits are such that this process although robust, produces parts out of specifications. In Fig 7.2d, the tolerances are wider than in Fig. 7.2c and therefore the same process now produces acceptable parts. Clearly, the process in Fig 7.2d is the most desirable process. Setting of the tolerance limits is done by the product engineer. In some cases, the engineer does have the flexibility to open up the tolerances

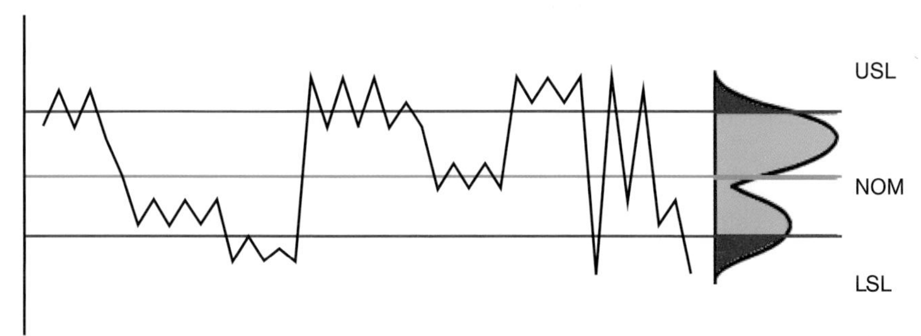

(a) Non-robust process with special cause variation producing parts out of specificatons.

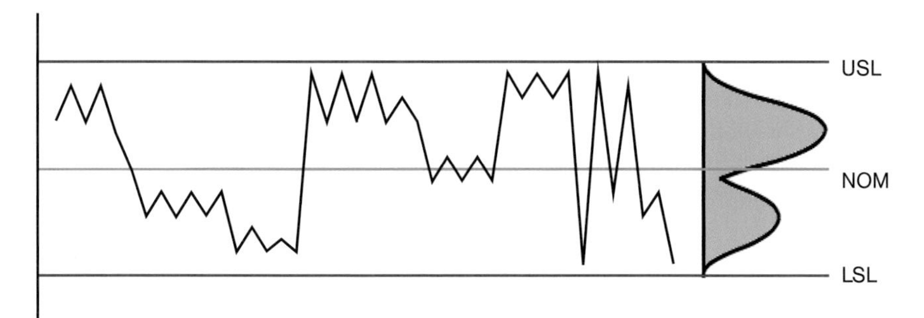

(b) Non-robust process with special cause variation producing parts within specificatons.

Figure 7.2a,b Types of processes based on variation and tolerances

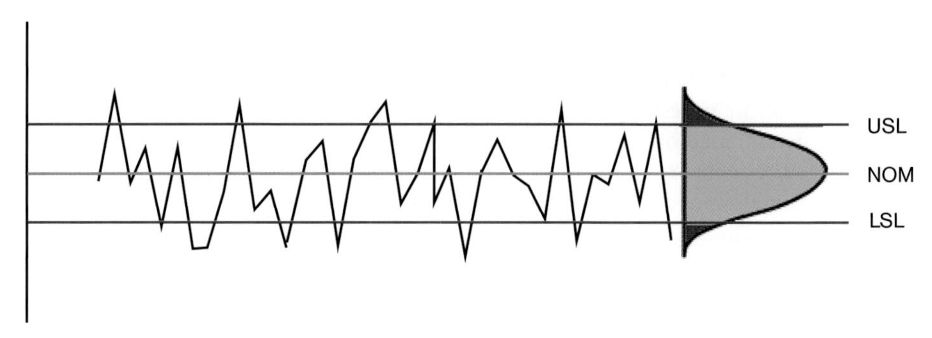

(c) Robust process with common cause variation producing parts out of specificatons.

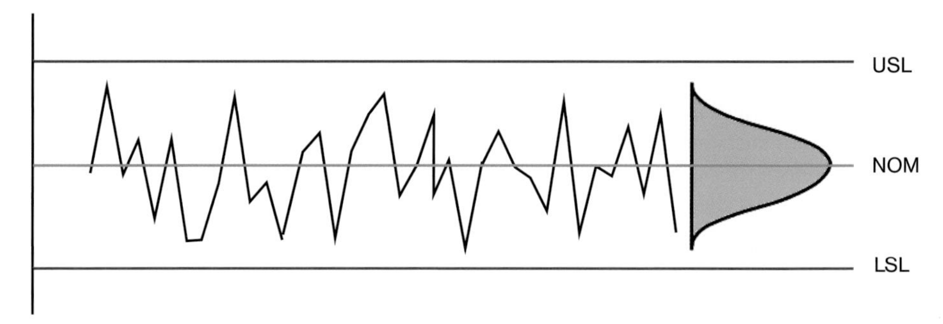

(d) Robust process with common cause variation producing parts within specificatons.

Figure 7.2c,d Types of processes based on variation and tolerances

based on the form, fit and function of the part. If opening of the tolerances is not acceptable to the product engineer and if it is not possible with the current process setup to reduce the variation, alternative solutions such as selecting another plastic, changing the amount of filler, or using cavity pressure control must be considered. As mentioned earlier, an attempt must always be made to eliminate the special cause variation, even if the parts are within the required specifications. This makes part quality more predictable and the manufacturing process less vulnerable to molding defective parts. Another benefit of implementing the discipline of developing robust processes is a reduction in part inspection frequency and sample sizes. The goal of process development, using scientific principles and techniques, must be to establish a process that is consistent and well within the specification limits similar to the one shown in Fig 7.2d. Simply molding parts within the specification limits does not necessarily mean that the process is robust and stable.

There are ways to improve the robustness of a process, reduce variation and improve consistency. The aim of process development should be to develop a stable and robust process.

There are systematic steps that must be followed in order to achieve this goal. Unfortunately, these steps are often ignored because they are time consuming and can increase the number of mold trial iterations. Often overlooked is the amount of time, energy, and materials that are wasted and scrapped due to production of non-quality parts: the mold needs constant attention of a technician to adjust the process to produce the parts within specification;

parts molded out of specification must be scrapped and re-run. This is a waste of material, machine time, and human resources. The time lost cannot be regained. Often, parts that are out of specifications are shipped to the customer where they will be discovered. This causes a loss of reputation, return of parts for rework, time consuming corrective action investigations, and the cost to scrap and remake the parts that cannot be reworked. With all the competition in a truly global market, efficiency in every area becomes an essential requirement. Consistency in output cavity-to-cavity, shot-to-shot and run-to-run are results of a robust process. Good molding practices start with a robust process development strategy.

7.2 Storage and Drying of Resin

Resin is supplied to the molder in a variety of ways. Most commodity resins, such as polyethylene, polypropylene, or ABS are typically used in very large volumes. The same material may be used in various products. Such resins are usually shipped to the molder in gaylords. For higher volumes, large silos are filled with resin which is brought in by trucks or even by rail cars. When the volume of material is not very large, such as for most engineering materials, the supplier usually provides resin in pre-packaged bags of 25 kg each. In some materials, a loss of properties can occur if the resin is not stored correctly. Some polyurethanes are examples of resins with shelf life limitations. Therefore, it is important to use this resin within the prescribed time specified by the manufacturer. In resins with shelf life limitations it is typically not the polymer that degrades, but rather the additives that can be time sensitive.

Other resins, such as acrylic, must not be stored such that they are exposed to sunlight because this may result in a loss of color or transparency. The resin must also be stored in a dry area to avoid unnecessary absorption of moisture. Some resin manufactures prepackage hygroscopic resins in their dry/processing state and ship them to the customer in sealed bags. Such resins do not need drying and can be used directly at the molding press. However, un-used resin that was stored back on the shelf must be re-dried before being used again. Clear identification of the resin and keeping a log of all the materials in use is considered good manufacturing practice. Many times, a sample resin is brought in, some of it is used and the rest stored for later re-use. A date, project ID, and material ID must be recorded on the sample bag to ensure the history of this material can be clearly identified.

7.2.1 Plastic Drying

The topic of plastic drying has been discussed in detail in Chapter 4. Hygroscopic resins must be dried in order to avoid loss of properties, surface blemishes, and/or internal defects. The loss of properties in some resins is a result of the hydrolytic degradation of the resin. In other resins, the presence of the moisture in the resin results in surface defects, such as splay or internal defects, such as voids. Processing difficulties, such as gassing and foaming of the melt, as in the case of nylons may be encountered. For these reasons, resins must be dried before processing. The specific amount of drying time and the correct drying temperature is dependent on the base polymer resin. The times and temperatures for each resin is dependent on the strength of the chemical bond the water forms with the polymer. For example, PBT needs to dry at 80 °C (180 °F) for 4 hours, whereas polycarbonates need to dry at 121 °C (250 °F) for about 3 hours.

Additives are often added to plastic to enhance their properties for specific applications and/or to reduce their cost. Fillers such as glass and minerals are the most widely used additives by volume. Other additives include plasticizers, lubricants, flame retardants, heat stabilizers, colorants, blowing agents, and biocides which are added to the polymers in small percentages. These are usually low molecular weight compounds and/or oligomers. If the drying time is extended beyond the manufacturers recommended limits, it is possible to degrade these additives and/or to cause them to leave the resin. For example, with PBTs and nylons it was found that if the suggested drying times were exceeded, there was a danger of producing defective or out of specification parts. Drying PBT resin for more than 8 to 10 hours produced brittle parts and loss in surface appearance. Parts that normally appeared black with high gloss now appeared dull grey. Figure 7.3 shows that PBT exhibited a drastic drop in the impact strength after it was dried for more than 12 hours [1].

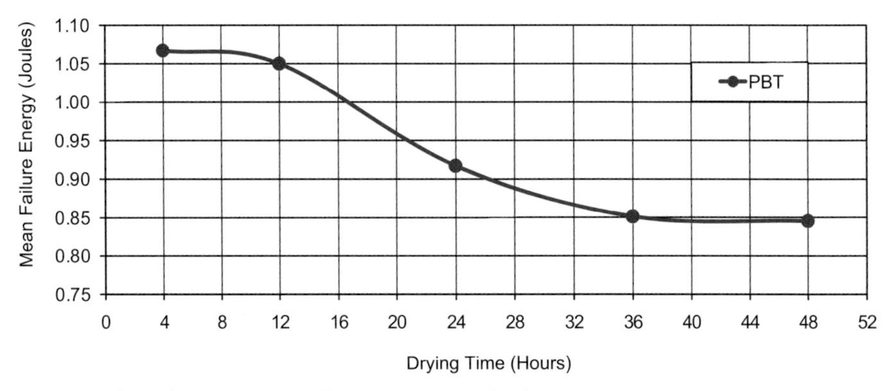

Figure 7.3 Effect of drying time on the impact strength of PBT [1]

Drying nylons for more than approx. 12 hours continually increased the viscosity of the plastic, causing constant adjustments to the molding process. The viscosity graph for nylon at different drying temperatures is shown in Figure 7.4 [1]. In this graph, nylon-4 represents nylon that has been dried for 4 hours and nylon-48 nylon that has been dried for 48 hours. Figure 7.5 shows the parts molded in the 'injection only' phase at various drying times. Parts molded within the 'injection only' phase are those that are molded with both pack and hold time and pack and hold pressure set to a zero value. Therefore the pack and hold phase are absent. With the same set process at the molding machine, as the drying time increases, the amount of part filling is less. This is a result of the increase in viscosity which reduces the flow of plastic. Overexposure or 'overdrying' must be prevented.

The most efficient ways to prevent over drying are:

- If the molding operation is not ready to be started after the material is dry, turn down the dryer temperature to about 80 °F, but keep the dryer running. Supplying the hopper with

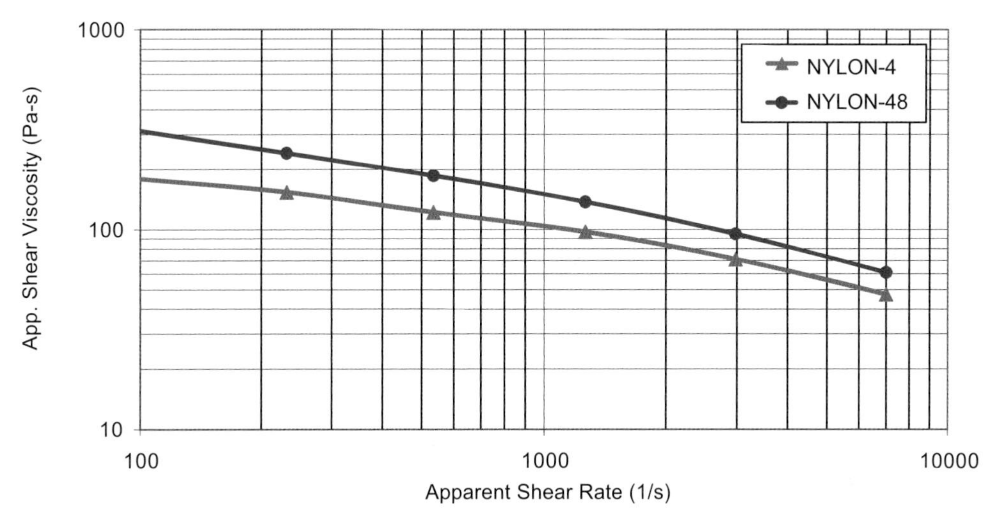

Figure 7.4 Viscosity of nylon dried for 4 and 48 h [1]

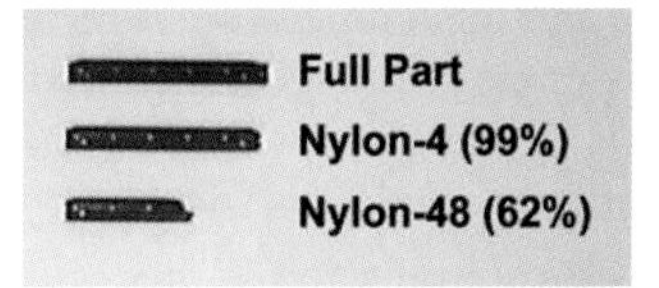

Figure 7.5 Parts molded using the 'injection phase only' with nylon dried for 4 hours and 48 hours

room temperature dry air will keep the moisture out and will not have detrimental effects on the resin.

- Size the hopper dryer such that the residence time of the plastic does not exceed the manufacturers' minimum/maximum recommended drying times for the plastic. When the available hopper sizes are larger than required, it is recommended to retrofit the dryers with an adjustable level sensor and adjust the sensor to load only the required amount of material.
- If the machine is going to be down for a long time (several shifts or days) after the drying has taken place, shut the hopper dryer off as soon as possible.

Precautions: Drying may have a cumulative effect in some resins, such as PBT. The drying process permanently removes the low molecular weight additives, which are not recovered when the drying is stopped and the resin is returned to the shelf for the next run. With nylons, moisture can be reabsorbed back into the resin. The effects of overdrying for PBT and nylon mentioned here must not be considered typical results for other materials. The effects of overdrying of other resins have yet to be studied because each resin and its additives are unique. Customized experimentation and testing will best suggest the type of drying control that is required. However, there is no real need to conduct this type of expensive and time consuming experimentation as long as the material is not exposed to excessive drying times.

7.3 Setting of the Melt Temperatures

Melt temperature selection should be the next step in the process development procedure once a material is ready to be processed or as the material is being dried. The material manufacturers provide processing data sheets that supply melt temperature information. In the case of amorphous materials, the range of recommended melt temperatures is wide, while for crystalline materials, the range is rather narrow, as was explained in Chapter 2. As a starting point, the mean of these recommended temperatures must be chosen as the target melt temperature. Note that the target melt temperature must be the actual melt temperature of the plastic and not the setting on the molding machine's barrel temperature controller. During the course of developing a molding process, the melt temperature will be varied and the final process may have a different melt temperature as a result of the process optimization study.

The temperature of the melt must be checked with the help of a melt pyrometer. A direct contact pyrometer is best suited for this purpose, because the probe is actually inserted directly into the melt to check the temperature. The 30/30 method of melt temperature measurement developed by RJG Associates is a commonly employed method. Here, the pyrometer probe is heated to a temperature of about 30 °F above the front zone temperature. The machine is then stopped on cycle, the nozzle retracted, and the melt purged. The heated probe is then immediately placed in the purge and the temperature is recorded after 30 seconds. The melt temperature must fall within the range of the recommended melt temperatures. In the case of crystalline materials, the melt temperature window is narrow, but in the case of amorphous materials, the melt temperature window is wide.

It is very important for the processor to refer to the processing data sheet before the start of molding. Two materials from the same chemical family can have completely different processing temperatures. For example, the OQ family of polycarbonate supplied by Sabic Innovative Plastics has a recommended melt temperature range of 305 to 332 °C (580 to 630 °F), whereas for the SP family of polycarbonate from the same company, the recommended temperature ranges from 248 to 271 °C (480 to 520 °F). Setting the temperature too high for the low melt temperature resin will degrade the resin and produce unacceptable parts. Setting the melt temperatures too low for the higher melt temperature resin may result in equipment damage, such as a broken screw or screw tip. Unfortunately, such failures are not uncommon on the production floor.

The settings for the machine barrel temperatures in order to achieve the right melt temperatures is different for amorphous and crystalline plastics. The barrel on a molding machine is divided into at least three heating zones and each of these zones can be set to a specific temperature. This results in a certain profile of barrel temperatures. As described in Chapter 6, the screw inside the barrel of the molding machine typically has three sections. Each section performs a particular function. The base of the screw is where the plastic pellets first come in contact with the screw. This part of the barrel is called the feed throat, it is designed to convey and then soften the pellets. The plastic must not melt here because if it did, it would stick to the barrel and/or the screw, preventing any additional material from conveying further. Therefore, the barrel temperatures closer to the feed throat must be set to the lower end of the recommended temperatures. The next section of the barrel is set at a higher temperature to start the melting process. The increase in the temperature of this zone depends on

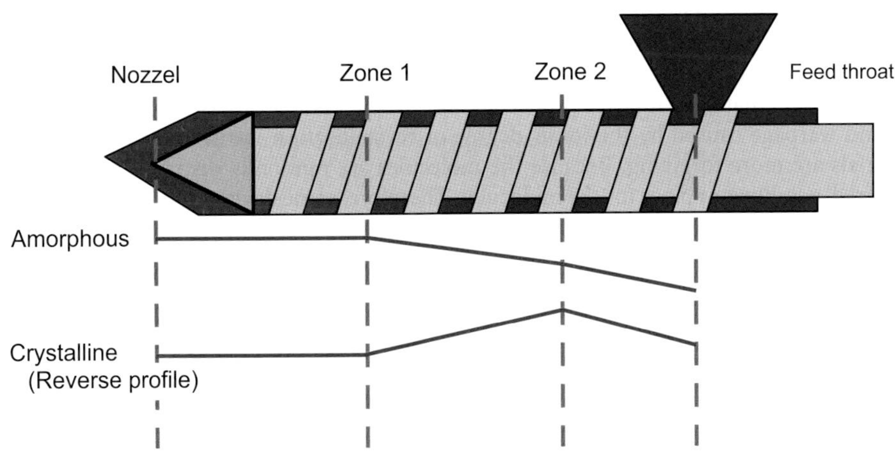

Figure 7.6 Barrel heat profiles for amorphous and crystalline plastics

the morphology of the plastic material. With crystalline polymers, the crystallites need considerable energy to soften and melt and therefore the increase in temperature is typically higher than for amorphous plastics. But since crystalline plastics can also be heat sensitive or cannot stand high temperatures for long periods of time, the temperature of the subsequent heating zones is reduced. This results in a heating profile that has a hump in the middle (also called a hump profile). In case of amorphous plastics, such a profile is not necessary because they need less energy to soften and can stand longer times in a heated barrel. The temperatures of the subsequent zones are higher and such a profile is called a conventional profile or regular profile. The different profiles are shown in Fig. 7.6.

A hump profile can also be used for materials that are more difficult to melt or when the material viscosity is high. Understanding the different sections of the screw and their functions as described in Chapter 6 will further help in establishing a setting for the melt temperatures. Typically, the nozzle temperatures are set within approx. 5 °C (or approx. 10 °F) of the desired melt temperature.

7.4 Setting Mold Temperatures

The processing data sheet provides the recommended mold temperature ranges that should be used. It is important to make sure that the actual mold temperatures are within this temperature range. The setting on a mold temperature unit (in case of water and oil units) is typically a few degrees higher because there is always a loss of heat that takes place during the transportation of the fluid. The actual mold temperature must therefore be measured. As discussed in the thermal transitions section of Chapter 2, the mold temperature is critical because it provides the energy for the molecules in the molten state to reach their final equilibrium resting states without creating molded-in stresses.

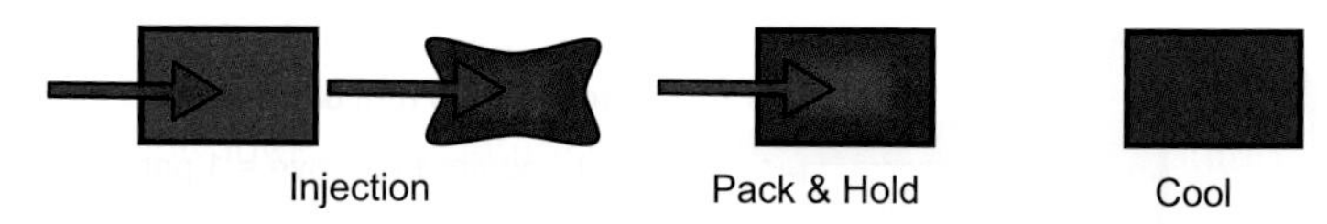

Figure 7.9 Representation of the injection, pack and hold, and the cooling phases

the plastic begins to cool, the molecules start to get closer to each other, resulting in shrinkage. At this point, if the addition of plastic were stopped, the parts would be under-packed and would exhibit a defect known as sink. Sink shows up as depressions on the surface of the part. Therefore, to avoid this and to compensate for the volumetric shrinkage, the injection phase is followed by the packing phase, where the required amount of plastic is packed into the mold. The required amount is equivalent to the volumetric shrinkage. At the end of this phase, the amount of plastic in the cavity must equal the theoretical weight of the part. The theoretical weight is equal to the solid density of the plastic multiplied by the volume of the cavity. Any less plastic in the cavity will result in an under-packed part and any more will result in an over-packed part.

The molten plastic enters the cavity through the gate. If this area freezes off, further filling of the cavity is impossible. Therefore, the gate must be large enough to prevent plastic freeze-off during the injection and pack phases. After the packing phase, the molten plastic is under very high plastic pressures, somewhere in the range of 35 to 70 MPa (5000 to 10000 psi), depending on the part. If the pressure on the screw is removed at the end of packing phase, the high pressure inside the cavity will force the melt out of the cavity. Therefore, a certain amount of pressure must continue to be applied in order to hold the melt inside the cavity. Care must also be taken not to apply excess pressure because that will pack in more than the required amount of plastic in the cavity and over-pack the part. This balance of pressure is called holding pressure and this phase of the injection process is called the holding phase. The holding pressure must be applied until the gate is frozen; this time is referred to as holding time. The gate must be frozen before the pressure behind the screw is released, otherwise the plastic inside the cavity will be forced out of the gate by the high pressure in the cavity. Figure 7.9 shows injection, packing, and holding phase.

It is difficult to determine the switchover point between pack and hold without employing some advanced methods, such as using cavity pressure measurement equipment. For the most part, pack and hold are taken as one phase and are usually called the holding phase. In many cases, the processes are set with profiles on the holding pressure to yield acceptable parts via trial and error. In doing so, the processor has actually set up a process that has a pack and a hold phase but is not aware of it or has not been able to distinguish the two phases from each other. Under-packing results in defects such as sinks and internal voids in the parts. Such parts usually exhibit some amount of post-molding shrinkage as well. Over-packed parts can have molded-in stresses that usually get relieved after the parts are ejected, resulting in defects such as warpage or premature failure.

7.5.4 Decoupled Molding^{SM}

The term Decoupled Molding^{SM} was coined by Rod Groleau of RJG Inc. (see Fig. 7.10). During the injection phase, the cavity is filled with molten plastic. The volume of melt injected dur-

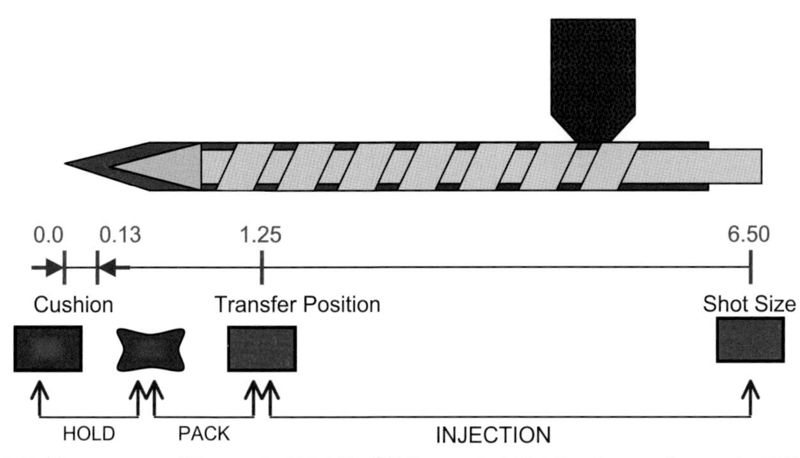

Figure 7.10 The concept of Decoupled Molding^{SM} (Decoupled Molding is a service mark of RJG Inc.)

ing the injection phase must be equal to the collective volume of the cavities and the runners. (Note that melt density is lower compared to solid densities.) Once this volume of plastic is present in the mold, the injection phase is followed by the packing phase which is then followed by the holding phase. Before such an understanding was developed by the molding industry, molders would either overpack or underpack the mold with molten plastic and would apply the pack and hold phase based on some previous experience. To stress the importance of the separation of the phases and educate the molders, the term Decoupled Molding^{SM} was introduced (see Fig. 7.10). Since each phase has a specific reason, they must be decoupled from each other and controlled separately.

According to RJG Inc., decoupled molding can be described by three different steps: Decoupled I, Decoupled II, and Decoupled III. In Decoupled I, the fill is accomplished via controlled velocity to a volumetric set-point before switchover to hold, where the plastic is held at a set pressure with no velocity. Melt inertia packs the mold. In Decoupled II, the mold is filled as quickly as possible to a set volume and the second stage hold is used to pack and hold the plastic in the mold. In Decoupled III, the mold is quickly filled to a set volume, packed to a cavity pressure set-point using a controlled velocity and then held until the gate is frozen. The phases are separated for understanding and control, but the screw is continuously moving during each of the three phases all of which are completed with one single stroke of the screw.

Decoupled molding provides the best control over the molding process and offers the most consistency. It is common practice to fill the cavity to 95–98% capacity during injection before entering the pack and hold phases because, in decoupled molding, the goal is to make sure the mold is not filled more than 100% during the injection phase. Targeting 95–98% provides a small margin of safety to make sure that the part is not excessively filled. The second reason is to compensate for the momentum of the screw during injection, which allows the melt to slow down before the start of the next phase. The packing phase is a less dynamic phase because we are only packing at the rate of shrinkage that is taking place in the cavity. A slowdown is therefore essential to insure the mold is not over-packed and or blown open and flashed.

7.5.5 Intensification Ratio (IR)

The concept of IR has been explained in detail in Chapter 6, see also Fig. 7.11. For hydraulic machines, the IR is the ratio of the cross sectional area of the hydraulic ram where the injection pressure is applied to the cross sectional area of the screw. The hydraulic pressure multiplied by the IR is equal to the plastic pressure present at the tip of the screw. For example, an IR of 10:1 means that a hydraulic pressure of 800 psi will provide a plastic pressure of 8000 psi at the tip of the nozzle. In other words, it is an amplification factor.

7.5.6 Screw Speed

The setting of the screw speed is another critical parameter, especially when it comes to crystalline materials. The shear from the rotating screw contributes a significant amount of energy to help melt the plastic. During the machine cycle, when the material is being processed through the barrel, the heat from the heater bands alone is not sufficient to melt the crystallites. The additional heat required is supplied by the shear energy created by the rotation of the screw. High screw speeds generate high shear and help in melting the crystallites.

Screw speeds must be set such that the screw recovery time is always less than the cooling times. The mold does not open until the screw recovery is completed and the cooling times are reached. If screw recovery takes longer than cooling time, the effective cooling time increases because the mold is still closed. In this case, the screw recovery time becomes the controlling parameter. The rule of thumb is to have the screw recovery completed approximately two seconds before the cooling time is reached. This is acceptable in some cases; however, it is not always practical. While molding crystalline resins, high screw speeds are required and, depending on the part, a longer cooling time may also be required. For example, the screw recovery may be completed in 4 seconds, but the required cooling time may be 10 seconds. In such cases, there is a significant delta between the two time intervals. Slowing down the screw speed will result in loss of the required shear and deteriorate the melt homogeneity. In such cases, the rule of thumb will not be applicable. In other cases, regardless of the morphology of the resin, where the cooling times are long, a very slow screw speed will also induce inconsistencies in the melt homogeneity or in the amount of the melt that is metered in by the screw for the next shot. Some amount of shear energy is always required to melt the plastic and improve the melt homogeneity. Looking at a cross section of the molding machine barrel with the plastic and the screw inside, the heater bands are on the outside, followed by the barrel, the plastic, and the screw. The plastic is a bad conductor of heat and therefore the inner layers of the plastic, closer to the screw, do not receive sufficient heat from the heater bands. The plastic is molten but the melt is not homogeneous. The shear from the rotating screw is what provides the required energy to melt these inner layers and contributes to melt homogeneity.

Recommended screw speeds are provided by the material suppliers. However, the values are typically given in revolutions per minute (rpm). This is very misleading because, depending on the diameter of the screw, a given value of rpm will result in high shear for large diameter screws while the same value will result in low shear for a smaller diameter screw. The screw

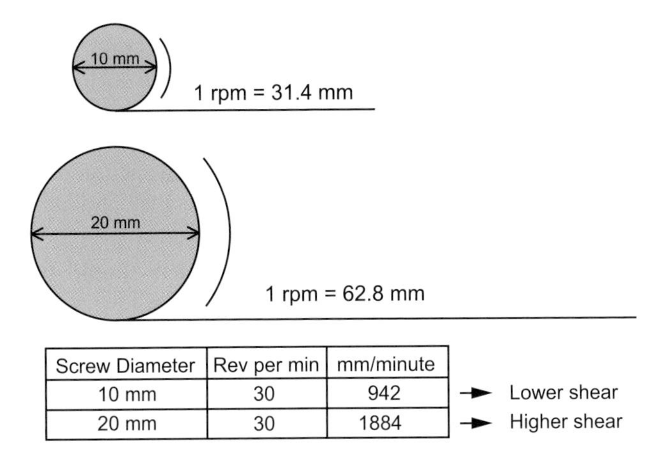

Screw Diameter	Rev per min	mm/minute
10 mm	30	942
20 mm	30	1884

Figure 7.12 Relationship between screw diameter, revolutions per minute, and surface linear speed

speeds should therefore be specified in distance per time units, such as inches per second or meters per second. An example is given in Fig. 7.12.

Highly filled plastics or plastics filled with long glass fibers typically require slower screw speeds to avoid a breakdown of the fibers. In these cases the screw design plays an important role in processing. Screw design also plays an important role in determining the melt homogeneity for different types of materials, depending on their shear sensitivity. In some cases, the cycle times can get longer because the available screw on the molding machine cannot generate enough shear heat to melt the plastic fast enough. Again specialty screws can make a difference here.

7.5.7 Back Pressure

The example of going down a ramp with a loaded cart illustrates the natural tendency to apply some force or resistance in the direction opposite to travel (see Fig. 7.13). This is done to better control the load and to gain consistency over the travel to avoid an accident. Similarly, as the screw moves back because of the buildup of plastic in front of the screw, an absence of pressure on the screw can cause the screw to move erratically. This loss of control

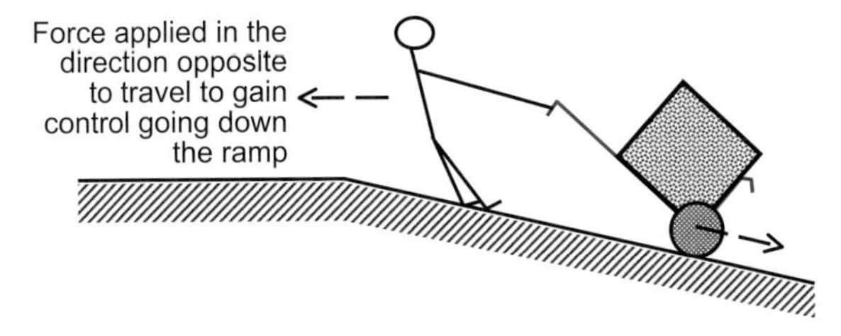

Figure 7.13 Explanation of back pressure for control and consistency

can lead to the screw reaching the shot size position inconsistently and prematurely, which in turn will result in process variations from shot to shot. In addition, the melt needs to be compacted in order to eliminate all gases and air that has built up during the screw recovery phase. Again, if the air and gasses are not eliminated, not only will the shot size be inconsistent, but the air and gasses will show up in the parts as voids, splay, or other surface or internal defects. Back pressure is also required to help in the mixing action of additives such as colorants. It is also essential in achieving good homogeneous melt and helps in the melting action of the pellets.

The back pressure values provided by the material suppliers should be used as guidelines only. The applied back pressure actually used should be as low as possible to achieve the required result. The consistency in screw recovery time is a good indication that the back pressure being applied is sufficient. The variation in screw recovery time must not exceed +/– 0.2 to 0.5 seconds on small to mid size machines and approx. 1 second on larger machines. Too much back pressure can cause excessive shear in the material which can degrade the material. Some fillers, such as glass fibers, can break down with excessive back pressure resulting in a loss of properties in the final product. In non shear-sensitive materials (but also in some shear sensitive materials) degradation of certain low molecular weight additives can result in the formation of gasses. This will increase the number of defects in the final parts and over time cause the vents in the mold to build up with residue and become clogged. Vents on the parting lines are easy to clean; however, the internal vent components, such as vent pins, are difficult to clean without pulling the mold from the machine and dismantling it. Excessive back pressure can also result in increased wear of barrel and screw components.

7.6 Process Optimization – The 6-Step Study

7.6.1 Step 1: Optimization of the Injection Phase –Rheology Study

All plastic melts are non-Newtonian. This means that their viscosity does not remain constant over a given range of shear rates. In the strict sense, the rheological behavior of a plastic is a combination of non-Newtonian and Newtonian behavior. At extremely low shear rates, which are rarely encountered in injection molding, the plastic is Newtonian; but as the shear rate increases, the plastic tends to exhibit non-Newtonian behavior. Interestingly, as the shear rates increase further, the plastic tends to act more and more Newtonian after an initial steep drop in viscosity. In Figure 7.14 the plastic viscosity is plotted on a log-log scale and in Figure 7.15 the plastic viscosity is plotted on the linear scale. Both graphs are for PBT and are plotted from identical data, the only difference being in the scale type.

On the linear scale graph it can be seen that the change or drop in viscosity is much greater at lower shear rates as compared to higher shear rates. This happens because with increasing shear rate, the polymer molecules start to untangle from each other and start to align themselves in the direction of flow. This reduces the resistance to flow (the viscosity). The plastic tends to get more Newtonian at higher shear rates. Although there is still a continuing drop

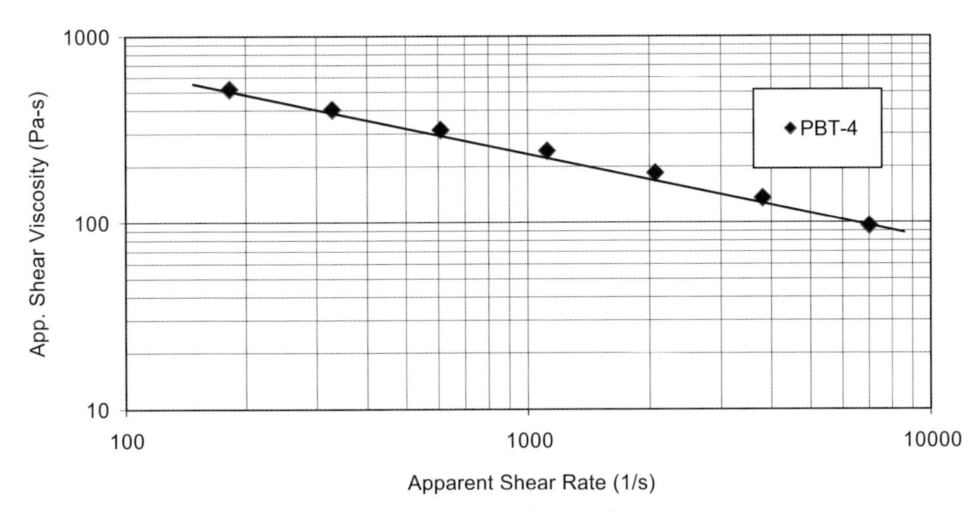

Figure 7.14 Viscosity of PBT represented on the logarithmic scale

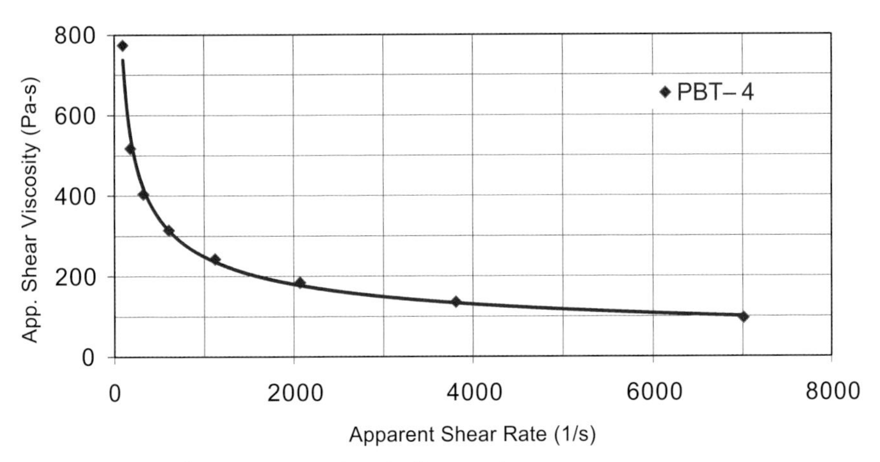

Figure 7.15 Viscosity of PBT represented on the linear scale

in the viscosity, the change is not as significant as at the lower shear rates. Figure 7.16 illustrates this phenomenon. For the sake of discussion, we shall refer to these regions as non-Newtonian and Newtonian regions. These are no truly Newtonian regions and the slope of the line depends on the nature and properties of the plastic. Typically, crystalline materials tend to have a flatter region compared to amorphous materials since the viscosities for crystalline materials are lower, facilitating the ease of orientation.

During injection molding, the material is subjected to high shear forces during the injection phase. The shear rate is proportional to the injection speed. If the shear rates are low and are set in the initial non-Newtonian region of the curve, then small variations in the shear rate

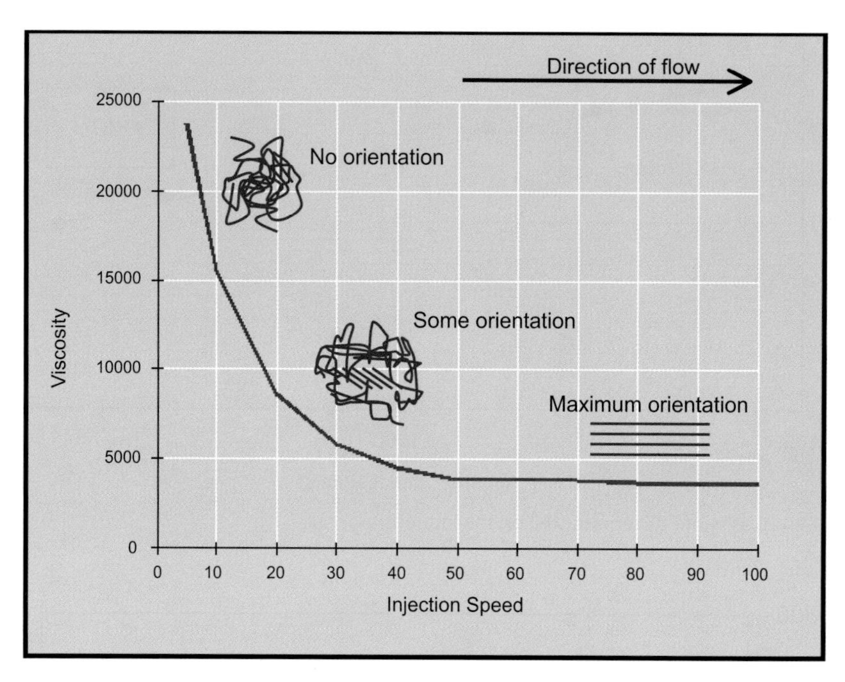

Figure 7.16 Orientation of the molecules in the direction of flow at different injection speeds (shear rates)

will cause a large shift in the viscosity. Since there is always some natural variation, the mold filling will be inconsistent and will therefore result in shot-to-shot inconsistencies. However, if the injection speeds are set to higher values, the viscosity tends to be consistent. At high injection speeds, the shear rates are high and the effect of shear rate on the viscosity is not as significant as it was at low injection speeds. Small changes in injection speed result in small or almost no change in the viscosity of the melt. Figure 7.17 illustrates this concept using data that was generated on a molding machine. Any natural variation in the speeds will not have a significant effect on the cavity fill in the Newtonian region of the curve and it is therefore important to find this region of the curve and set the injection speed, and with that the shear rate, here.

The viscosity curve can be generated at the molding machine for a given mold. This is known as 'In-Mold Rheology Study' or simply 'Developing the viscosity curve'. A study by Mertes et al. [2] demonstrated that at high speeds the variations in the end of fill pressure were smaller compared to the variations at low speeds. Perturbations were introduced on purpose and two polypropylenes with high and low viscosities were used in the study. The results are shown in Fig. 7.18.

The other advantage of using high injection speeds is the reduction of the effects of lot-to-lot variations. Figure 7.19 is a compilation of viscosity tests performed on various lots of a certain grade of polycarbonate supplied to a molder on a specific project. Different lots of the same material can have different viscosities at the same injection speed. However, the difference is greater at the lower injection speeds.

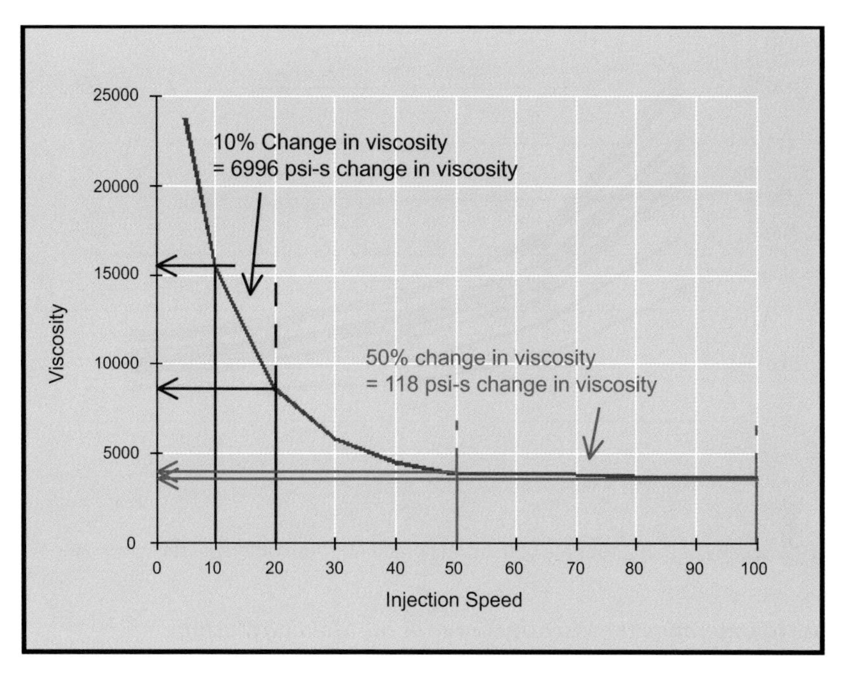

Figure 7.17 Effect of change in injection speed on the viscosity of the plastic in the non-Newtonian and Newtonian region

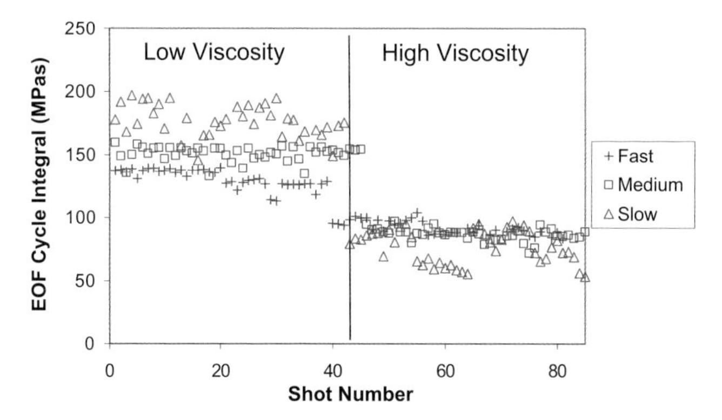

Figure 7.18 End of fill cycle integral data at various injection speeds for two polypropylenes of different viscosities [2]

The procedure to determine the viscosity curve was first developed by John Bozzelli. The basic principle is based on the melt rheometer that is used to study the viscosity of plastics (see Chapter 3). The injection molding machine can also be treated as a rheometer, where the nozzle orifice is the die of the rheometer and the screw is the piston. The hydraulic pressure is applied to the molten plastic with the help of the screw. The pressure required to move the screw at a set speed is recorded.

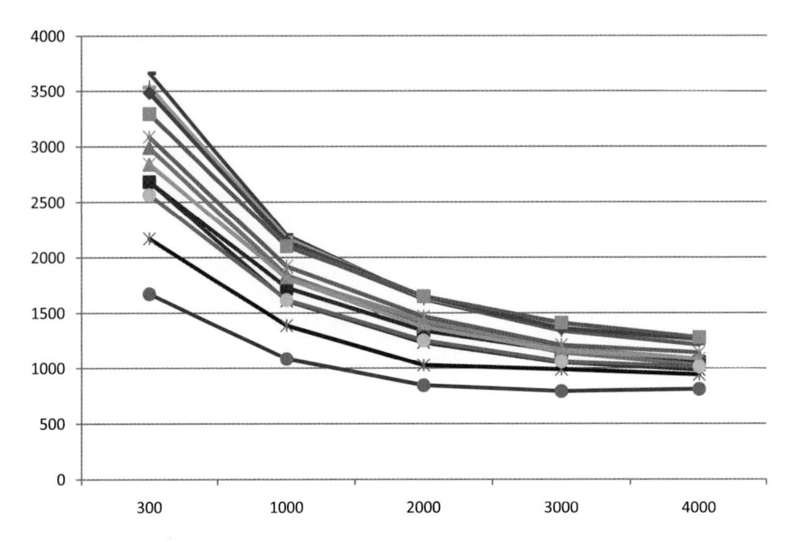

Figure 7.19 Viscosity of polycarbonate of the same grade but from different lots

Procedure to Determine the Viscosity Curve at the Molding Machine

Safety should always be given first priority when conducting any experiments.

- Set the melt temperature to the one recommended by the manufacturer. If there is a range, set the temperature to the center of the range.
- Set all the holding phase parameters to zero. This means that there will not be any holding phase but rather only the injection phase.
- Provide a screw recovery delay time equal to the approximate time that would allow the gate to freeze off. This would be based on some past experience. For example, if for a similar part and material the gate freeze-off time was 5 seconds, set the screw delay time to about 8 seconds. This must be done so that there is enough time given for the material to freeze off in the mold before the screw back pressure is applied during recovery. If the material is not sufficiently cooled in the gate area, the cavity can fill because of the back pressure and the exact amount of material injected during the injection phase will not be known.
- Set the injection pressure to the maximum available value. There is an assumption here that the mold and the machine are compatible in terms of shot size and plastic pressure usage. If the shot used is small or if the required pressure is considerably less than the maximum available, then set the pressure to lower values to avoid any mold damage or over-packing of the cavity.
- Set the cooling time to a safe value such that the part will be cool and has reached the ejection temperature before mold opening.
- The machine must be set to transfer on screw position. Set the injection speed to 'slow' and make a part. The part should be short. If not, adjust the transfer position to make the part such that it is filled only to about 50 %.

- Incrementally increase the speed as close to the maximum as possible in small steps and make sure that the parts are still short. If the part is full, adjust the transfer position, such that the part is about 95 – 98 % full by volume. This means that at close to the maximum possible injection speed the parts are 95 – 98 % full with no holding time or pressure. If this is a multi-cavity mold, the part that fills the most must be the one that is 95 – 98 % full.
- Make another shot and record the fill time and the peak hydraulic pressure at transfer. For example, the machine is set to 2200 psi but may require only 1850 psi to move the screw at the maximum speed of 5 in/s. If the hydraulic pressure was set to a value other than the maximum available and during the experiment the peak pressure matched the maximum value, the set pressure must be increased. The process must not be pressure limited.
- Next, lower the speed by a small amount, for example from 5 in/s to 4.5 in/s or from 90 % to 80 %. Note the fill time and the peak injection pressure.
- Repeat the above steps until you achieve the lowest injection speed possible. Divide the range of available injection speed into about 10 – 12 speeds so that you get the most data points possible.
- Obtain the intensification ratio of the screw from the machine manufacturer. If this number is not available, use a value of 10. It is important to understand that the intensification ratio is necessary to calculate the plastic pressure. However, this should not be a show stopper for the experiment or the optimization process. Since this is a constant used in the equation, the graph will shift up or down respectively, but the profile of the graph will remain the same, which is of primary interest to us.
- To get the viscosity, use the following formula:

  Viscosity = Peak Injection Pressure × Fill Time × Screw Intensification Ratio

Plot the graph of viscosity versus injection speed. Figure 7.17 shows a typical viscosity curve generated at the molding machine. Table 7.1 shows a sample worksheet that can be used for the study.

Table 7.1 Viscosity Curve Worksheet (Screw Intensification Ratio = 10.5)

No	Inj. speed (in/s)	Fill time (s)	Peak hyd. press (psi)	Viscosity (psi-s)
1	0.20	5.75	735	42263
2	0.50	2.37	805	19079
3	1.00	1.28	963	12326
4	1.50	0.87	1112	9674
5	2.00	0.68	1240	8432
6	2.50	0.57	1339	7632

Table 7.1 *(continued)* Viscosity Curve Worksheet (Screw Intensification Ratio = 10.5)

No	Inj. speed (in/s)	Fill time (s)	Peak hyd. press (psi)	Viscosity (psi-s)
7	3.00	0.50	1427	7135
8	3.50	0.44	1504	6618
9	4.00	0.41	1588	6511
10	4.50	0.38	1663	6319
11	5.00	0.36	1669	6008

How to Use this Information

Looking at the created curve, it is noticeable that the viscosity stays fairly constant after about 50 % of the injection speed. Therefore, setting the injection speed to 60 % would ensure that the filling stage of the process will stay consistent. Any small natural variations will not cause large changes in viscosities resulting in shot-to-shot variations as it would at lower injection speeds. The injection speed must be selected closer to the "knee" part of the curve. This is where a shift in the viscosity and greater consistency can be observed. It is not advisable to inject the plastic as fast as possible for a couple of reasons. First, plastic materials can be shear sensitive and will degrade at higher shear rates or injection speeds. Second, it is always difficult to get the best venting possible in a mold because of the very valid concern of flashing plastic into the mold vents. For this reason mold makers are very conservative with vent depths and reliefs. A mold maker will typically machine in the smallest vent clearances possible, as vents are not a steel safe condition. Venting is also sometimes difficult in areas such as deep pockets or corners that are not on the parting line of the mold. A slower injection speed helps in such cases. Optimizing the injection speed through in-mold rheology is the first step to achieving a robust process.

Cautions and Exceptions

Generating the viscosity curve is a great tool to find out the effect of injection speed on the viscosity of the plastic. However it must not be used in all cases and there are exceptions.

Insert molded components typically have inserts that need to be partially or wholly encapsulated with the plastic and in most cases, these parts are held inside the cavity. Injection of plastic at high speeds can easily dislocate and/or deform the insert, thus molding an unacceptable part. High speeds also tend to shear the material and in some cases cause gate blush or splay. Optical components can sometimes exhibit this problem, which necessitates that lower speeds be used. In addition, at higher speeds, gas must be vented from the mold very efficiently. Venting some sections of the mold can become tricky and difficult. In such cases, only the fastest possible speed without burning should be used. In these molds, the vents must be cleaned very frequently because any build-up will result in burning of the plastic.

Another practice that should be followed is to slow down the speed of injection before going into the second phase. This will ensure an accurate and repeatable switch-over point. If the screw is not slowed down before entering the holding phase, the momentum of the screw can influence the momentum of the melt causing inconsistencies. It is similar to applying the brakes before coming to a stop sign or entering the garage.

In some molds it is not possible to continually make short shots because the parts can stay in the fixed side of the mold. In this case, an attempt must be made to find the 95–98% full cut-off position at close to the maximum injection speed. There are a couple ways this can be done. First would be the method described earlier and would involve removing the part every time the short shot occurs. This will dictate that the machine be run in a semi-automatic mode. The second possible procedure would consider the volume of the melt required to fill the complete cavity, equate this to the volume of the injection cylinder, calculate the required shot size and start the procedure by using about 95% of the shot size. In this procedure it is acceptable to add any holding pressure to fill the part so that it pulls away from the fixed side of the mold and stays on the ejection side during mold opening. The two values that are required in the calculation of the viscosity are the peak injection pressure and the fill time. These two parameters comprise the injection phase and can be obtained from the machine manufacturer. Therefore, what happens in the holding phase, which follows the injection phase, does not affect these values. Whether or not the parts get filled and packed out is not of concern. Not having any holding pressure or time and producing a short shot has its value; however, in cases where this is not possible, it is acceptable to fill the parts.

When observing the short shots from a viscosity study, it is evident that the parts get shorter and shorter as the injection speed is lowered. The question that usually comes to mind here is why are the parts getting smaller if the shot size and transfer position is not changed?

There are two phenomena explaining this condition. First, the viscosity of the plastic is being reduced with the decreased injection speed. One must remember that when the viscosity of the melt is lowered, its flow rate also becomes lower. Second, the momentum of the screw and the melt also reduces with decreasing injection speed. When the hydraulic pressure is abruptly cut off at the end of the injection phase, the screw and/or the melt can still travel further and fill the cavity due to the inertia of the screw. At higher injection speeds the momentum is higher and the viscosity is lower.

The position of the screw can also be tracked by recording the furthest forward position it reaches during the injection phase. The cushion value displayed on the machine can be misleading because this number represents the position of the screw at the end of the holding phase and may not necessarily represent the furthest position of screw travel. In some cases a screw bounce-back can cause the cushion value to be higher. In case of electric machines there is no hydraulic pressure but instead a servo motor that controls the position of the screw. The servo system will stop the screw instantly with almost no momentum but the melt will still have some momentum. Pictures of a fill progression during a viscosity study series are shown in Figure 7.20.

As the injection speed increases, the fill time will decrease and the shear rate will increase. However, the pressure required to maintain the injection speed is an aggregate of the individual pressures of the various phenomena taking place as the plastic melt travels to the end of fill. Therefore with increasing injection speeds, the required pressure can either increase

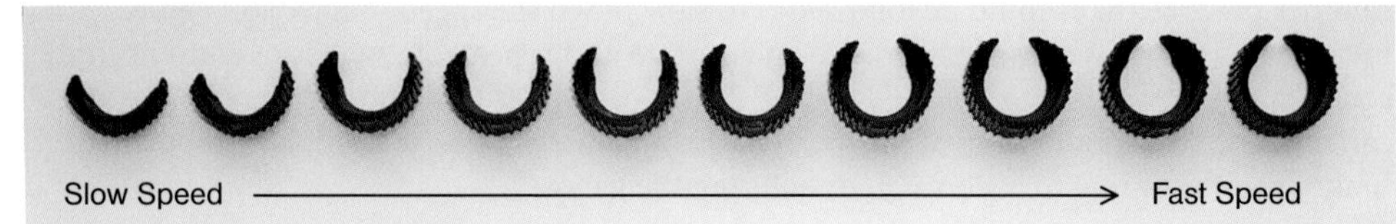

Figure 7.20 Progression of fill during a in-mold rheology study from a slow to fast injection speed

or decrease, depending on the dominant phenomenon. These phenomena and the resulting required pressures are described in the following.

- As the melt hits the mold, it starts to cool, resulting in an increase in the viscosity of the plastic. This will result in an increase in the required pressure.
- The flow channels in the mold typically call for progressively smaller cross sectional areas towards the end of the flow. Smaller cross sections result in higher required pressures. However, if the cross sections are generous, the resulting pressure may not change.
- When the plastic first enters the mold, the layer of plastic next to the wall forms a frozen layer and the hot melt that is entering the mold now flows in between these frozen layers. This is also called fountain flow. This frozen layer gets thicker and thicker during the filling phase, again increasing the required pressure to move the screw at the set injection speed. With thick parts or generous flow channels this effect is not significant because the filling may be completed before the frozen layer is sufficiently built up.
- Because plastics exhibit non-Newtonian behavior, with increasing injection speeds the viscosity drops and the pressure required to move the screw will reduce.

7.6.2 Step 2: Determining the Cavity Balance – Cavity Balance Study

A specific volume versus temperature graph was discussed in Chapter 2. The graph is shown in Fig. 7.21 with the areas that correspond to the different phases of the injection molding cycle. The injection phase corresponds to the plastic being in the melt phase and as the melt starts to cool, the pack and hold phase come into effect. When the pack and hold is done, the plastic is cooled down below the ejection temperature where it can be safely ejected out of the mold.

Shrinkage is directly related to specific volume. As the pressure increases, the specific volume decreases. The packing pressure therefore affects the shrinkage of the part. The higher the packing pressure, the higher is the pressure of the plastic inside the cavity and the lower is the shrinkage. Therefore, to control the dimensions of a part, the packing pressure or cavity pressure must be controlled. Incidentally, this is the principle behind the use of cavity pressure sensing technologies that will be described in Chapter 12.

Figure 7.22 shows the effect of cavity pressure on the length of a tensile bar. The tolerances are also shown on the graph. As the cavity pressure increases, the length of the tensile bar increases.

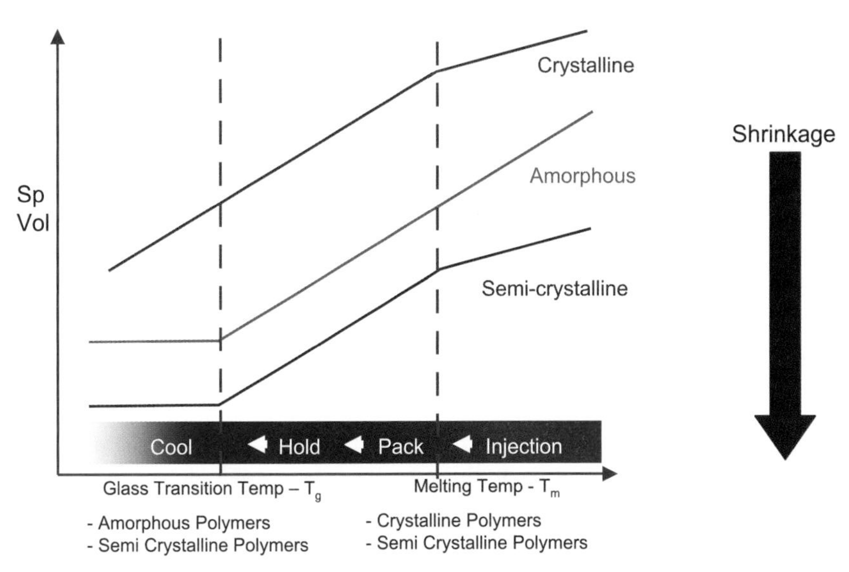

Figure 7.21 Effect of temperature on specific volume and the application to injection molding

Now let us consider a two-cavity tensile bar mold. To mold tensile bars of equal length, the plastic pressure in each of these cavities must be identical. When the cavity fill is identical, the pack and the hold phases that follow will have the same effect and produce identical cavity pressures. If one cavity happens to fill less than the other, then at process switch-over from pack to hold, the cavity that fills first will get pressurized more, leading to higher cavity pressure and therefore a larger dimension. This will result in cavity-to-cavity dimensional variation. One cavity may replicate the first tensile bar dimension in the figure and the other may replicate the last one. When one is within specifications, the other may be out of speci-

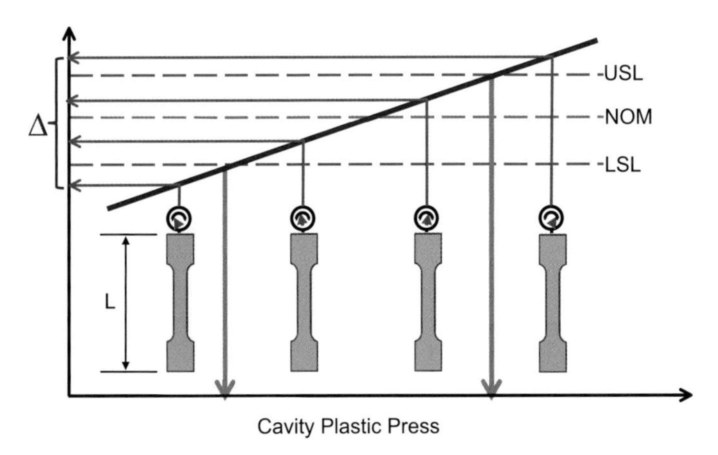

Figure 7.22 Effect of cavity pressure on the length of a tensile bar

fications. In addition, when the process windows are small, the cavity that fills first may end up with flash, even when the other cavity is still unfilled (short).

Figure 7.23 shows parts from a two-cavity mold, where one part is short while the other exhibits flash. To avoid these issues, the flow of the melt into each of the cavities must be identical. Determination of cavity balance is therefore an important step during process development.

There are several reasons for flow imbalances:

- *Flow channel variations:* If the runners and the gates are not identical, there is more restriction to flow in one cavity than in the other. This will lead to a fill imbalance.
- *Venting variations:* Vents are added to the mold in order to provide a pathway for air in the cavity to escape as it is being displaced by the plastic melt that is entering the cavity. Therefore, the vents must be capable of allowing air to escape from the cavity at a rate equal to the flow rate of the plastic into the cavity. Consider a worst case situation, where one cavity does not have any vents. In this case, the plastic would not be able to enter the cavity as fast as it could in the other cavity. There would be some amount of cavity fill due to natural venting, but it would be not nearly as effective as in the other cavity with the deliberate addition of vents. Therefore, if the size of the vents and the position of the vents are not identical, this will affect the fill pattern leading to dimensional variations in the parts.
- *Cooling variations:* The mold temperature can significantly affect the flow of the plastic, especially with crystalline materials and parts with long flow lengths. Generally, higher mold temperatures result in higher flow rates of the plastic. Mold temperature will also affect the amount of shrinkage. Therefore, if the cooling is not identical in all cavities, the plastic will flow and shrink differently in each cavity, causing dimensional variations. A similar effect is seen when a water line is plugged or not connected correctly, causing one section of the mold to run hotter than another section where coolant is flowing properly.
- *Rheological flow variations:* This topic was discussed in detail in Chapter 3. Consider the following example: in an 8-cavity mold with an H-shaped runner, the four inside cavities

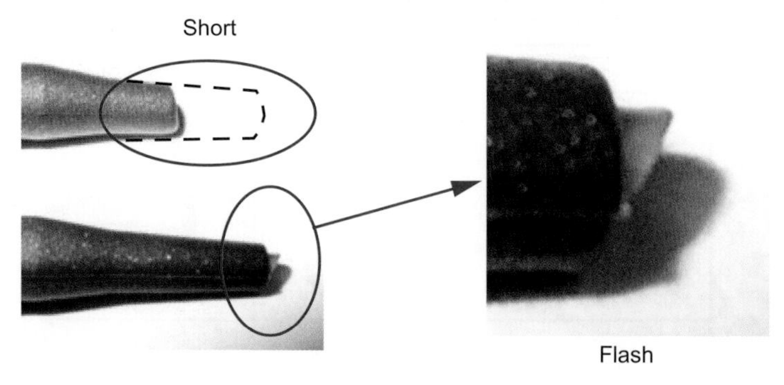

Figure 7.23 Two-cavity mold with one cavity short and the second with flash molded in the same shot

will fill more than the outside cavities because there is a flow imbalance that results from the melt rheological imbalance even if the runner is geometrically balanced. To produce identical parts in all 8 cavities, it is imperative that the flow in the two pairs of inner cavities is balanced with the two pairs of outer cavities. Because of the inherent imbalance in an H-pattern 8 cavity mold, it is difficult to balance all eight cavities from a rheological perspective.

Procedure to Determine the Cavity Balance

In order to achieve cavity balance it is important to determine the reason(s) for any existing imbalance. First, the actual steel temperature in each cavity must be checked via a reliable method. Contact type probes work best. A thermal imaging system can also be very useful because it provides a quick picture even while the mold is running. Any difference in the cavity temperatures should be close (within a temperature difference of less than 2 °C or 5 °F).

Next, the incoming inspection of the mold should include a record of the runner and gate sizes which must be identical. If no such record exists, the runner and gates sizes must be inspected and recorded. Runners, and especially gates, wear over time. Tracking these data over time is very valuable. These two factors will determine if cooling and flow channel variations could pose a problem. The procedure outlined in the following will determine if any flow imbalance exists because of rheological or venting factors.

- Set the holding pressure to zero.
- Set the holding time to zero.
- Set the screw recovery delay time to a value close to an estimated holding time, similar to the step in generating the viscosity curve in the previous section.
- Set the cooling time to a value that will ensure that the part will be cool enough to eject.
- Set the injection speed to the value obtained from the viscosity curve study.
- With the rest of the settings identical to the viscosity study, start molding. Because this test is done immediately after the viscosity study, the settings must yield a short part. If there is a visible cavity imbalance, the cavity that fills the most of all cavities must still be unfilled. For example, in a four-cavity mold, if cavity number 4 fills first, then in this study start molding parts so that cavity 4 is about 98 % filled. At this setting, the other cavities will now have filled less than the 98 %. All cavities will now be underfilled. Save the short shot and record the weights of each cavity. Take an average of at least two shots.
- Next, generate a short shot series and record the cavity weights for each shot as in the step above. Depending on the size of the part, at least 3 to 4 progressing shots must be recorded, starting with the smallest short shot parts that are possible to make. Next, the data must be plotted on a chart similar to the one shown in Fig. 7.24.

Table 7.2 shows an example of the worksheet that can be used to document the cavity balance study.

Table 7.2 Cavity Balance Worksheet

Cavity ID	Part weight (g)				
	10% Part	25% Part	50% Part	75% Part	End of fill
1	1.94	3.15	6.12	8.34	12.94
2	1.92	3.14	6.20	8.21	12.85
3	1.95	3.25	6.92	8.86	12.12
4	1.95	3.36	6.55	8.52	12.82

How to Use the Cavity Balance Information

Figure 7.24 shows that at lower percentage fills, all the cavities filled evenly. However, as the fill percentage increased, cavities 6 and 7 began to fill less than the others. This effect was caused by the fact that the venting for these two cavities was not identical to the venting of the rest of the cavities. In cavities 6 and 7 vents were present but they were not as deep as the others. This prevented the air from getting out as fast as it needed to, which slowed down the fill of the plastic.

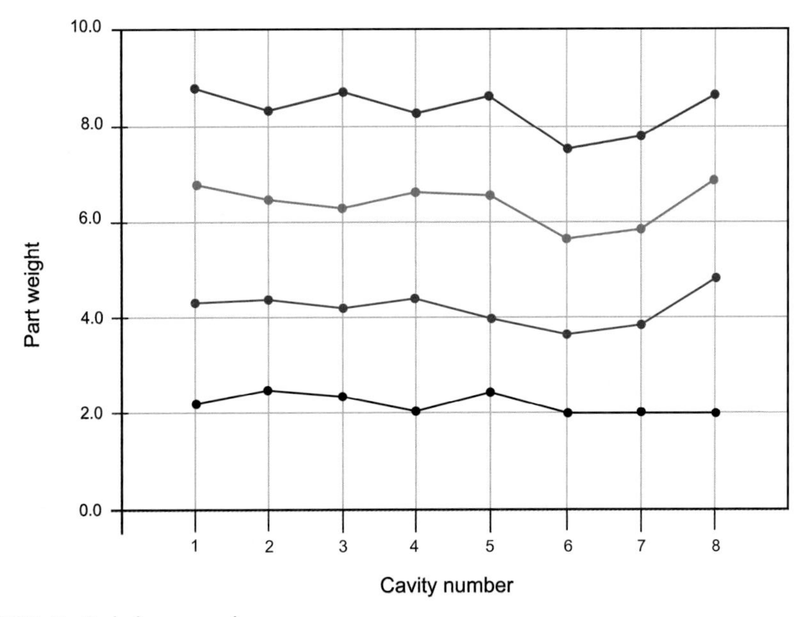

Figure 7.24 Cavity balance graph

These cavity imbalances must be taken care of to achieve effective processing that will produce consistent parts within specifications. First, the vents must be checked and made identical with the rest. Variations in vent sizing can be considered a steel variation and should be checked as part of a new mold inspection procedure or routine preventative maintenance procedure.

Another advantage of performing a short-shot series is that it will indicate whether the gate sizes are the same. In the example shown in Figure 7.24, initially all cavities filled identically, indicating hat the gate sizes were the same.

Rheological imbalances can be fixed by altering the melt flow based on the principles and procedures developed by Beaumont Inc. Before the introduction of this technology, it was common to open the flow channels or the gates in order to balance out the flow. When cavity balance is discussed, it is not just the fill pattern that we are interested in. Ultimately, the quality of the melt that is delivered to the different cavities must be identical. The quality is defined not only by the flow rate, but also by the pressure and temperature history of the melt. Although modifying the runners and gates will balance out the flow rate, it will not balance out the pressures or the temperatures.

The acceptable percentage variation depends on a variety of factors, the most important ones being the dimensional tolerances for the part and the available process window. A wider process window provides better chances of packing out all cavities. Although packed out, as described above, the cavity-to-cavity dimensions will be different because of the differences in cavity pressures. However, if the tolerances are wide enough and all parts are falling within specifications, the percentage variation can be considered acceptable. A wider tolerance and wider process window provide for a higher acceptable percentage of variation. Conversely, tighter tolerances result in a lower percentage of acceptable cavity-to-cavity variation. The final quality of the parts should be checked to see if there is a need to tighten up the balance between the cavities. For example, a Cpk value derived from a population that consists of all parts from all cavities is a good measure.

The polymer morphology also plays a role here. Amorphous materials can tolerate more imbalance than crystalline materials. Amorphous materials do not shrink as much as crystalline materials and therefore cavity-to-cavity dimensional variations are smaller. At the same time, because crystalline plastics flow easier, they tend to flash easily, reducing the process window. Balancing out the flows between cavities helps improve the process window regardless of the morphology of the material and whether the tolerances are wide or not.

Family Molds

Family molds are molds that do not mold the same part in all cavities. It is common to build family molds for small production runs of assemblies. The balance of fill is important in family molds, too. Again, the concept here is the same as before, where the cavity pressure is the deciding factor for the dimensions. Typically in these cases, the gate and runner sizes are adjusted to balance out the fill. Depending on the configuration, the number of cavities and the cavity layout, a decision must be made on the type of cavity fill balance technique.

7.6.3 Step 3: Determining the Pressure Drop – Pressure Drop Studies

As plastic flows through the different sections of the nozzle and the mold, the flow front of the plastic experiences a loss of applied pressure because of drag and frictional effects. Additionally, as the plastic hits the walls of the mold, it begins to cool, increasing the viscosity of the plastic, which in turn requires additional pressure to push the plastic. The skin of plastic that is formed on the internal walls of the runner system decreases the cross sectional area of the plastic flow, which also results in a pressure drop. Depending on the pump capacity of the molding machine, there is a limited maximum amount of pressure available to push the screw at the set injection speed. The required pressure to push the screw at the set injection speed should never be more than the maximum available pressure. If the pressure required is higher, the screw will never be able to maintain the set injection speed throughout the injection phase and the process is considered pressure limited. Initially the set speed may be reached, but as soon as the process becomes pressure limited, the screw slows down as shown in Fig. 7.25.

During process development, knowing the pressure loss in every section of the flow path helps to determine the overall pressure loss and the sections where the pressure drops are high. The runner system of the mold can then be modified to reduce this pressure drop and achieve a better consistent flow. Sudden changes in pressure drop from one channel into another are also not desirable. The mold must be modified to reduce this pressure drop, if the process is pressure limited. The sections where there are sudden changes in the pressure must also be modified to achieve a better consistent flow.

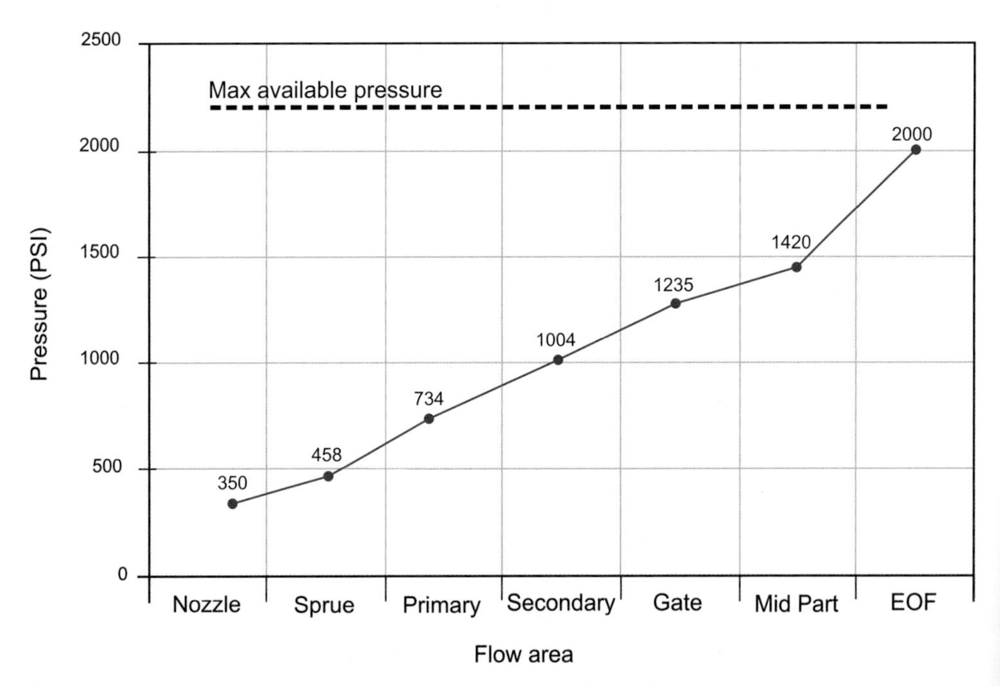

Figure 7.25 Graph to determine the pressure drop through the mold

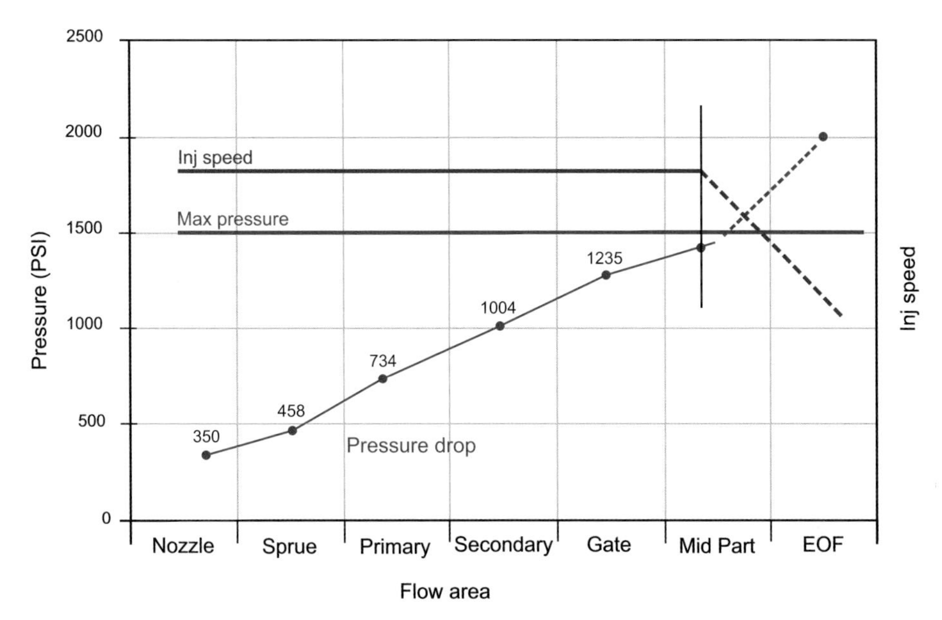

Figure 7.26 Effect of a pressure limited process on injection speed

Procedure to Determine Pressure Drop

Consider the plastic flows through the following sections: the nozzle of the machine, the sprue, the primary runner, the secondary runner, the gate, and the end of fill. The procedure to determine the pressure drop is as follows.

1. Set the machine process temperatures as described in Sec. 7.3 and set the injection speed according to the viscosity curve study. Set the holding pressures and times to zero and set the screw recovery delay time using the same procedure followed during the viscosity study.
2. Set the machine to the maximum available pressure.
3. Build a shot on the molding machine and take an air shot. Note down the peak pressure required to do so.
4. Next, start molding but adjust the shot using the transfer position only and mold only the sprue while recording the peak pressure.
5. Next, adjust the transfer position only and mold the sprue and the primary runner only while recording the peak pressure.
6. Repeat step 5 and mold up to the end of the secondary runner only while recording the peak pressure.
7. Repeat such that the plastic just enters the gate and record the peak pressure.
8. Repeat such that the plastic fills 1/3rd and 2/3rd of the part and record the peak pressures.
9. Repeat such that it just reaches the end of fill and record the peak pressure.
10. Plot a graph of the peak pressure versus the flow section to generate the pressure drop graph.

Figure 7.27 shows a typical graph generated from a pressure drop study. The study can also be done by reversing the steps taken. Start the study with a 95 – 98 % part and work your way slowly backwards towards the sprue. Naturally one would know right away whether the mold is pressure limited but the study will indicate the section with the highest pressure drop. In some cases, molding just the sprue can create problems because it may get stuck inside the mold. In such cases, stopping the study at the end of the primary runner is acceptable. However, in all instances the pressure drop through the nozzle should always be measured and recorded. The pressure drop through the nozzle is inversely related to the size of the orifice. As the orifice size decreases, the pressure drop increases. During mold startup, the actual pressure drop through the nozzle must be compared to the recorded pressure drop. This will give an indication whether the right size and type of nozzle is installed on the machine. For example, the nylon type nozzle will have a slightly higher pressure drop compared to a conventional nozzle for the same diameter orifice.

If any problems arise with the part filling during production, one of the first actions must be to pull the barrel back and check the pressure drop through the nozzle. Any debris, such as metal shavings stuck in the nozzle, can cause an increase in pressure drop. This loss of the pressure drop at the nozzle gets added to the sprue, the runner, the gate, and the part. If the process is close to being pressure limited, this can easily lead to short shots and defective parts. Moreover, the loss in pressure in the runner system leads to a loss in cavity pressure and therefore a change in dimensions, shown in Figure 7.27. The type and orifice size of the nozzle are therefore very important and must be recorded as part of the setup sheet.

In hot runner molds the procedure should start with the nozzle and then continue with the section where the melt is first injected into the mold. The hot runner section of the mold must be considered as one flow section and then each subsequent flow section must

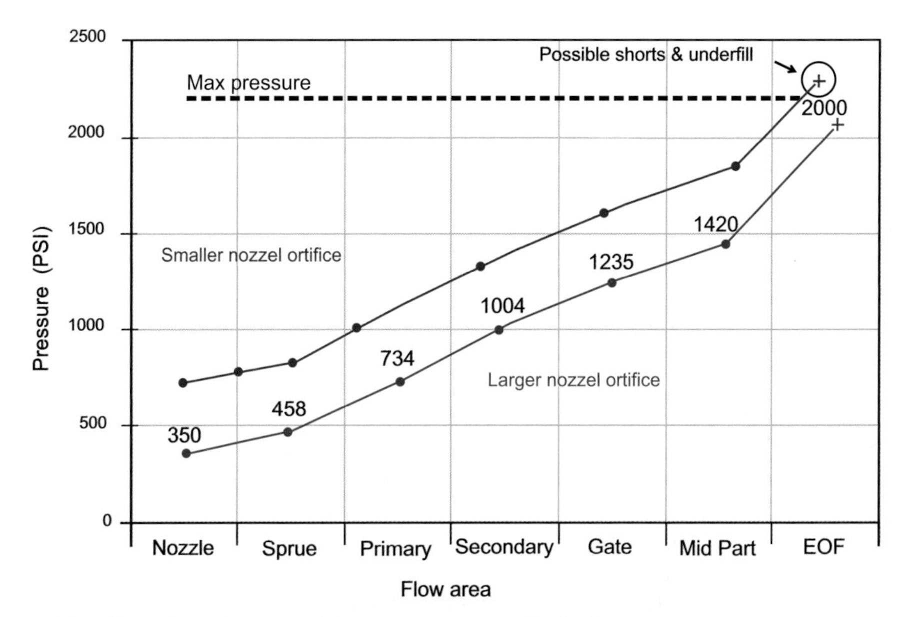

Figure 7.27 Effect of nozzle type and size on near pressure limited processes

be analyzed. If it is possible to manually inject the plastic at the set speed, an injection of the plastic into an open mold should be performed and the pressure drop should be recorded. For safety reasons some machines do not allow the processor to inject plastic at the full set injection pressure. In such cases, the data should still be recorded and the machine should be clearly identified on the setup sheet. Safety should be of primary concern and no safety features of the machine should be compromised. In the case of a hot runner mold, injecting through the manifold with the mold open can provide insight into the pressure drop through the hot runner system. Preferably, a piece of thick cardboard should cover the moveable side of the mold to prevent any splatter of plastic onto these surfaces. A cardboard shield should also be placed underneath the bottom of the mold for the purge and another over the top of the mold to prevent any unexpected burst of high pressure plastic from splattering above the mold and falling onto personnel near the machine.

The pressure drop through the different sections of the part must also be evaluated. Thin sections in a mold require more pressure to fill and can make a process pressure limited. Although the runners and the gates are generously designed, the part can have restrictions that may cause a pressure drop.

Table 7.3 is an example of a worksheet that can be used to document a pressure drop study.

Table 7.3 Pressure Drop Study Worksheet

Section	Peak pressure (psi)
Nozzle	236
Sprue	454
Primary runner	734
Secondary runner	1004
Tertiary runner	1248
10 % part	1630
50 % part	1837
End of fill	2030

How to Use this Information

The maximum pressure used in the process should never reach the maximum available pressure on the particular machine. For example, if the maximum available hydraulic pressure is 2200 psi, the end of fill pressure should not reach 2200 psi. Typically, only up to 90 % of the maximum available pressure should be used. In the generated graph, if the process is pressure limited or reaches more than 90 % of the maximum, steep increases in the required pressure between two sections must be identified. For example, if the required pressure to fill through the nozzle, sprue, primary and the secondary runner is 900 psi, and if through the tertiary runner the pressure increases to 1600 psi, this increase is significant compared to the increases in the previous sections. Modifications to the tertiary runner must be made. Increasing the diameter of the runner will help in reducing the pressure.

7.6.4 Step 4: Determining the Process Window – Process Window Study

As described earlier, the injection of plastic into the cavity can be divided into two main phases. The first phase is the injection phase where the mold cavity is completely filled with the molten plastic. The volume of the melt is equal to the volume of the cavity. The second phase is the packing and holding phase. The holding pressure must pack additional plastic material, equivalent to the volumetric shrinkage caused by cooling as the plastic makes contact with the cold mold walls. The various parameters that need to be controlled during this phase are packing pressure, holding pressure, packing time, and holding time. In most cases, packing and holding are not differentiated and are collectively called the holding phase.

The packing phase consists of packing the cavity with plastic equal to the theoretical weight of the part. The theoretical weight is equal to the solid density of the plastic multiplied by the volume of the part. Any less plastic will result in an under-packed part while any more will result in a part that is over-packed. Under-packed parts show defects such as sinks and internal voids. Such parts usually exhibit considerable post molding shrinkage. Over-packed parts can have built-in stresses that usually get relieved after molding resulting in defects such as warpage or premature mechanical failure. The ideal pack and hold pressure is deter-

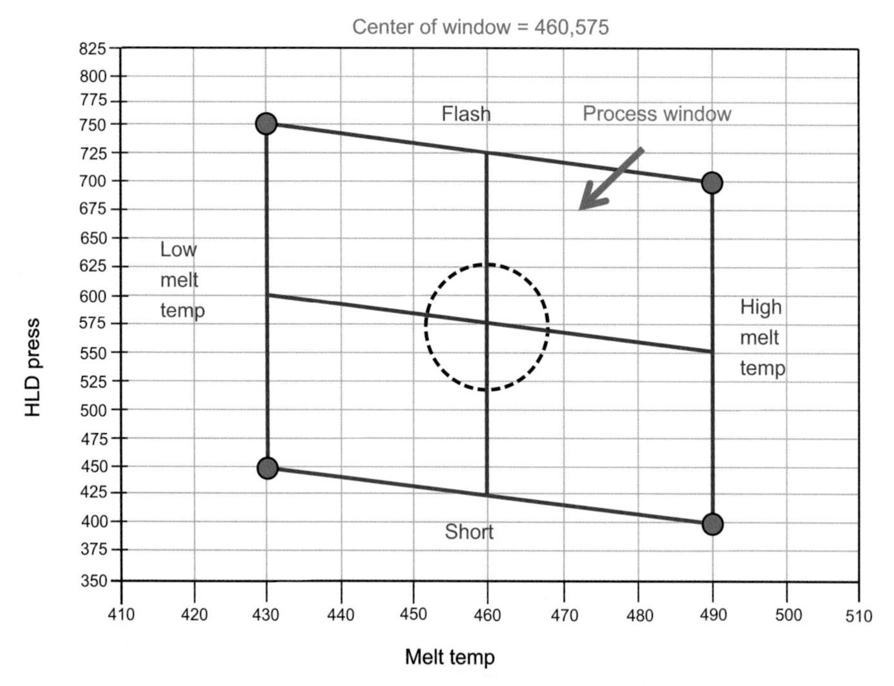

Figure 7.28 Process window study for amorphous materials – hold pressure vs. melt temperature

mined by evaluating the process window of the mold. Two process variables are varied to establish the process window. For amorphous materials, the holding pressure and the melt temperatures are the common variables that are used. For crystalline materials, because here the melt temperature range is narrow, the melt temperature is fixed to a medium value of the recommended values and the two variables that are selected are holding pressure and mold temperature. Holding pressure usually has the most impact on the quality of the part because it directly influences the specific volume and therefore part dimensions. Therefore, the significance of holding pressure applies to both types of materials. For amorphous materials, melt temperature usually has the higher impact compared to mold temperature. For crystalline materials, where the degree of crystallinity depends on the mold temperature, the two variables chosen are holding pressure and mold temperature.

The process window is also called the molding area diagram. This is the area in which aesthetically acceptable parts are molded. The bigger the window, the more robust is your process. It must be noted that the process window provides only the range of processing parameters in which aesthetically acceptable parts are produced, also called the aesthetic process window (APW). Chapter 9 is dedicated to the different types of process windows.

Dimensional data must be taken and the mold steel must be adjusted to move the dimensions to the center of the window in order for the process to be robust. Figures 7.28 and 7.29 show the process windows for amorphous and crystalline materials, respectively.

Outside the illustrated process windows, parts produced will not be acceptable because of defects such as sink, flash, or built-in stresses. Below the melt temperature limits, un-melted

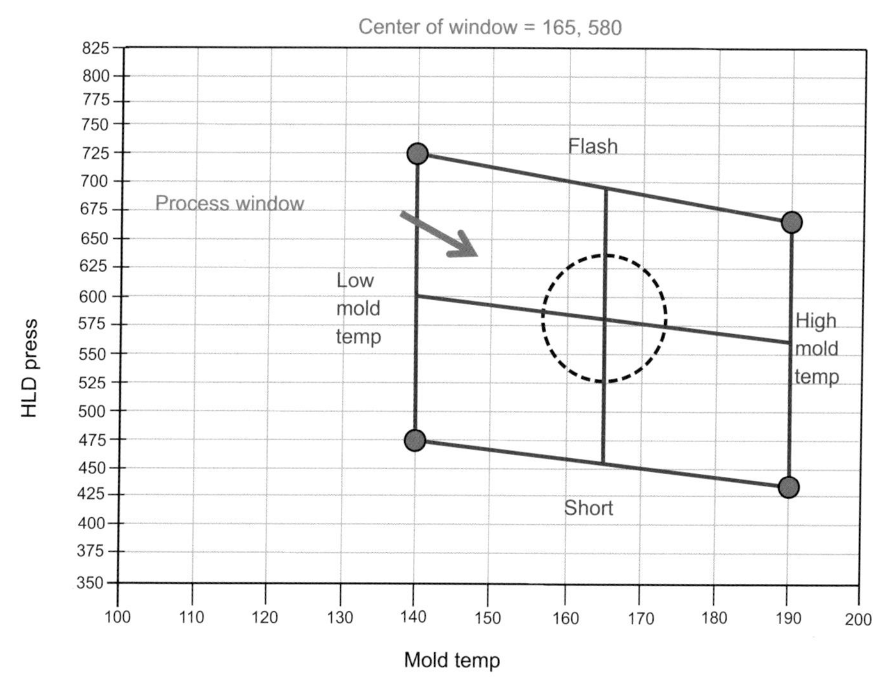

Figure 7.29 Process window study for crystalline materials – hold pressure vs. mold temperature

plastic may cause problems while above the temperature limit degraded plastic may render the parts unacceptable. With low mold temperatures the plastic may not have enough energy for crystallite formation. On the other hand, with mold temperatures higher than necessary for crystallization, cycle time may increase. The process is set to the center of this window so that any variations within the window would still yield cosmetically acceptable parts. The larger the window, the more robust is the process.

Chapter 8 will define and further discuss the concepts of dimensional process window (DPW) and control process windows (CPW).

Procedure for Determining the Process Window for Amorphous Materials

The following details the procedure for developing a process window for amorphous materials using holding pressure and melt temperature as the two variables.

- Set the mold temperatures to a medium value within the range of recommended mold temperatures.
- Set the barrel temperatures to attain the lower value of the recommended melt temperatures.
- Set the injection speed to the value obtained from the viscosity curve experiment.
- Set all holding times and pressures to zero.
- Set the screw delay time to a value as described in the viscosity study experiments, a number approximately equal to the estimated holding time for the part.
- Set the cooling time to a value higher than what would be typically necessary (e.g., if the estimated cooling time is 10 seconds, set the cooling time to 20 seconds).
- Start molding and adjust the transfer position to make a part 95 – 98 % full.
- Let the process and the melt stabilize by molding approx. 10 shots of parts.
- Now set the hold time to a value that makes sure the gate is frozen (in the next section we will learn how to optimize this time). The determination of hold time is typically based on previous experience. For example, for a 30 % glass filled PBT or nylon with a gate size of 0.070", the hold time typically ranges between 6 and 10 seconds. For this part of the experiment, we would set the hold time between 10 to 12 seconds.
- Increase the holding pressure in small increments and record the pressure at which the first cosmetically acceptable part is made. There should be no shorts, sinks, or voids.
- Record this pressure as the 'Low Temperature – Low Pressure' corner.
- Increase the pressure further in similar increments and record the pressure at which there is evidence of an unacceptable result, such as part sticking in the mold, flash on the part, or warpage. Record this pressure as the 'Low Temperature – High Pressure' corner.
- Repeat the last two steps, but this time at the high end of the recommended melt temperature range. This time the two extreme parameter combinations would be 'High Temperature – Low Pressure' and 'High Temperature – High Pressure' corners.
- Joining these four corners will generate the process window or the molding area diagram.
- Set the process to the center of this window.

Procedure for Determining the Process Window for Crystalline Materials

The following is the procedure for developing a process window for crystalline materials using holding pressure and mold temperature as the two variables.

- Set the melt temperature to a medium value in the range of recommended melt temperatures.
- Set the mold temperatures to attain the lower value of the recommended mold temperature range.
- Set the injection speed to the value obtained from the viscosity curve experiment.
- Set all holding times and pressures to zero.
- Set the screw delay time to a value as described in the viscosity study experiments, a number approximately equal to the estimated holding time for the part.
- Set the cooling time to a value higher than what would be typically necessary (e.g., if the estimated cooling time is 10 seconds, set the cooling time to 20 seconds).
- Start molding and adjust the transfer position to make a part 95–98% full.
- Let the process and the melt stabilize by molding the approx. 10 shots of parts.
- Now set the hold time to a value that makes sure the gate is frozen based on previous experience.
- Increase the holding pressure in small increments and record the pressure at which the first cosmetically acceptable part is made. There should be no shorts, sinks, or voids.
- Record this pressure as the 'Low Temperature – Low Pressure' corner.
- Increase the pressure further in similar increments and record the pressure at which there is evidence of an unacceptable result, such as the part sticking in the mold, flash on the part, or warpage. Record this pressure as the 'Low Temperature – High Pressure' corner.
- Repeat the last two steps, but use the high end of the recommended mold temperature. This time the two extreme parameter combinations would be the 'High Temperature – Low Pressure' and 'High Temperature – High Pressure' corners.
- Joining these four corners will now generate the process window or the molding area diagram.
- Set the process to the center of this window.

Table 7.4 is an example of a worksheet that can be used to document the process window study.

Table 7.4 Process Window Worksheet

Melt temperature (°F)	Low hold pressure (psi)	High hold pressure (psi)
430	550	1050
510	600	1100

How to Use this Information

The size of the process window is an indicator of how much variation the process will tolerate while still producing cosmetically acceptable parts. The aim is to have a wide process window. If the process window is very narrow, there is always a danger of molding parts with defects. For example, the graphs generated via the process window studies show that with a very small process window even natural process variations could cause occasional short shots or flash. A robust process is one that has a large process window and accommodates the natural variation inherent in the system.

A mold that produces short parts and parts with flash at the same time does not have any process window. The mold must be fixed because it will be impossible to make acceptable parts. Molds with extremely small process windows usually have dimensional issues. The process at which the parts are cosmetically acceptable produces parts that are out of specifications. When a certain specification is achieved, the parts are usually not cosmetically acceptable. Achieving consistent production from such molds is a difficult task and the processor often has to continually adjust the process. A few molds with small process windows on the production floor can easily consume available resources and make a plant run inefficiently. Determination of the process window is the most important process engineering study of all and if nothing else, this is the one study that must be done.

Cautions and Exceptions

The process window described in this chapter is the cosmetic process window. Within this processing window parts can be molded that are cosmetically acceptable. The procedure calls out for increasing a certain process parameter, such as the holding pressure, to a limit where the parts would have a cosmetic defect such as flash or the parts stick in the mold due to over packing. However, consider a case where the parts do not flash even at high holding pressures or do not show any signs of over packing. In these cases, although there is no evidence of failure, the parts could very well be over packed. In such cases, the higher pressure at which cosmetically acceptable parts can be molded must not be taken as the upper limits of the process. An evaluation of the parts and the high pressure limits is required and the process must then be set accordingly. In such cases, cosmetic process windows can be misleading because part failures may not be seen until they are used in an assembly.

7.6.5 Step 5: Determining the Gate Seal Time – Gate Seal Study

The molten plastic enters the cavity through the gate. The mold filling phase is dynamic, during which melt temperature, pressure, and flow velocity are all changing with time. For the mold fill phase, time begins with injection, which is the start of the forward movement of the screw. As the cavity begins to fill and is nearly full, the pack and hold pressure phase starts. The melt flow velocity is reduced and the melt temperature simultaneously drops. This causes an increase in viscosity of the melt. The gate has a fixed cross sectional area. When the viscosity of the plastic in this and the surrounding area it drops to a value at which the plastic cannot flow anymore, the gate is considered frozen. The plastic molecules in the gate area are now immobile and cannot flow into the cavity anymore. The time it takes to reach this stage is called the gate freeze time.

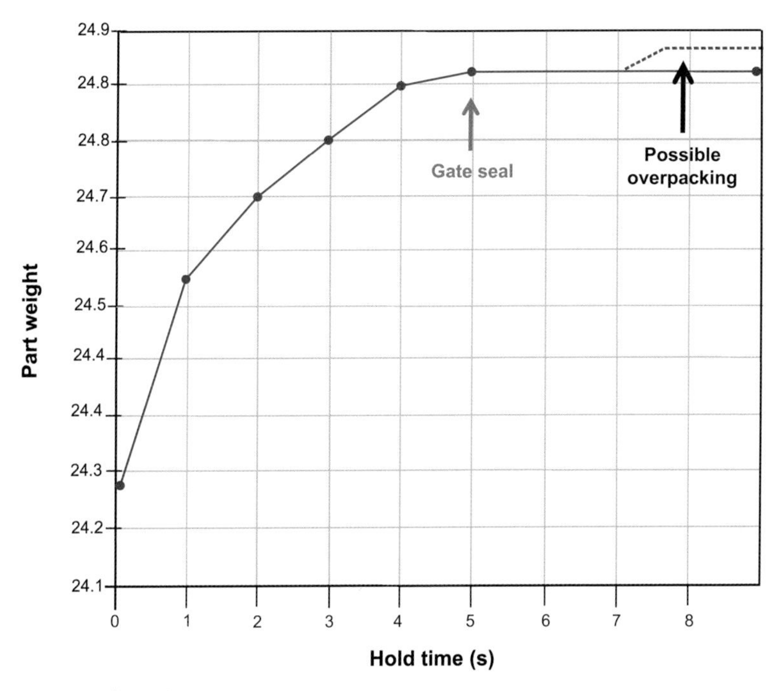

Figure 7.30 Gate seal graph

For an injection molding process, pressure must be applied to the melt until such time that the gate is frozen. If pressure is not applied for a sufficiently long time, either the part will be under-packed resulting in internal voids or sinks; or plastic pressure inside the cavity is high enough to flow back out of the cavity, which will also result in under-packing. The second phenomenon usually takes place when the holding time is just a little shorter than the gate freeze time, while the cavity is full of pressurized plastic. Gate freeze time is a function of the type of plastic, gate size, gate design, and the processing parameters of the machine. A gate freeze study must be completed for every mold. A gate freeze study is a graph of part weight versus holding time. Once the gate is frozen, the part weight stays constant because plastic can no longer get into or out of the cavity. A constant part weight is an indication of gate freeze. Figure 7.30 shows a typical gate seal graph.

Procedure to Determine Hold Time

- Set the injection speed to the value obtained from the viscosity curve experiment.
- Set the process at the center of the process window or towards the upper corner of the right hand side upper quadrant of the process window generated in the previous section. Holding pressure, melt temperature, and mold temperature values may influence the gate seal time. The higher the pressures and temperatures, the easier is the movement of the plastic into and out of the cavity. Higher temperatures will also increase the time it will take to freeze the gate. The aim should be to make sure that anywhere in this pro-

cess window the gate must always be sealed and therefore conditions towards the higher extremes of the pressures and temperatures must be considered.

- Set the cooling time to a value to ensure that the part is cooled before ejection.
- Drop the holding time to zero and again set the screw delay time as described in the viscosity study.
- Start molding approximately 10 shots.
- Increase the holding time to one second and collect a shot.
- Increase the holding time in increments of one second and collect the shots up to a holding time value at which the gate should be frozen.
- Weigh the shots and plot a graph of part weight versus time.
- Determine the gate seal time. The gate seal time is the time after which the part weight stays constant.

Table 7.5 is an example of a worksheet that can be used to document a gate seal study.

Table 7.5 Gate Seal Study Worksheet

Hold time (s)	Part weight (g)
1	12.98
2	13.26
3	13.84
4	14.06
5	14.12
6	14.23
7	14.23
8	14.23

How to Use this Information

The hold time must be set to approx. 1 second longer than the gate seal time. This will ensure that the gate will always be sealed before the end of the holding phase. The set cooling time starts at the end of the holding time. The plastic begins to cool as soon as it hits the walls of the mold. Therefore, actual cooling time is the sum of the injection fill time, pack and hold time, and the cooling time. This total cooling time is one of the factors that define the quality of the part. If the holding time is reduced by 0.5 seconds, this time must be added to the cooling time and vice versa. It is therefore best to add 1 or 1.5 seconds to the hold time to ensure gate freeze and then reduce the cooling time by the same amount, keeping the total cooling time the same. In an attempt to reduce cycle time, some processors add only about 0.5 seconds to the gate seal time. This may not be the most optimum setting.

Cautions and Exceptions

With a gate seal study the time at which the gate seals off or is frozen is determined. However, if the part needs a large gate, the seal-off time can be very long. In such cases, if the pressure is applied for longer than the required time, there is the risk of over-packing the part. Sprue-gated components are good examples of this problem. In such cases, the parts must be evaluated to find out the ideal pack and hold time. If the part weight stays constant during an initial section of the holding time and if with a further increase in holding time there is a sudden increase in the part weight, this can be evidence that the part is being over packed, especially in the gate area. Because the plastic behind the gate and in front of the gate is probably still soft, the extra time causes the frozen gate section to dislocate, forcing more plastic into the cavity. Since the rest of the plastic in the cavity is cooling down and viscosity is very high, the excess pressure from the fresh plastic does not get distributed or compensated for and builds up stress in the gate area. This weight increase is shown with the dotted line in Fig. 7.3. In materials such as low density polyethylenes, excessive times may force the cold plastic into the cavity, increasing the part weight. This problem may also be caused by short flow lengths from the sprue to the gate, which facilitates easier pressure transfer.

7.6.6 Step 6: Determining the Cooling Time – Cooling Time Study

The plastic starts to cool down as soon as it hits the walls of the mold. Once the holding time is over, the set cooling time counter starts. The mold remains closed until the end of the cooling time. The mold then opens and the part is ejected. Before the mold opens, the part must reach the ejection temperature of the plastic. If the part is ejected before it reaches ejection temperature, the part is too soft and will get deformed during ejection. Warpage can also be an issue. Excessive cooling time is a waste of machine time and therefore profits. Cooling time should also be set so that the part dimensions remain consistent and the process is capable of molding acceptable parts over time. The procedure to determine the optimum cooling time is relatively simple and involves molding parts at various cooling times and measuring their dimensions. The procedure is outlined in the following.

Procedure to Determine Cooling Time

- Set the process with the process conditions determined by the viscosity study, process window study, and the gate seal study.
- Mold three shots at various cooling times.
- Measure the critical dimensions.
- Plot a graph of dimension versus the cooling time.
- Analyze the data to see how the critical dimensions are changing with the cooling time.
- Decide on a cooling time that best fits the data.
- Run 30 shots at this cooling time and perform a statistical analysis to determine the process capability at this cooling time.

Determination of the right cooling time can become complicated. For parts with thick sections it is difficult to measure the internal temperature in the center of the thickest section. In

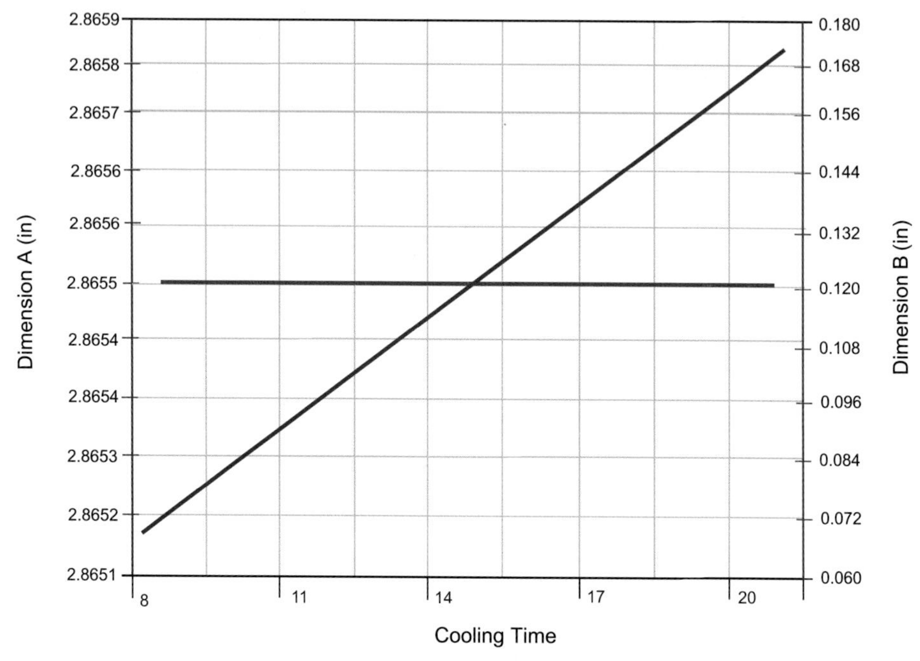

Figure 7.31 Cooling time study graph

some parts of the mold, it is difficult to get enough cooling and therefore cooling times have to be increased to allow time for additional heat transfer. In some cases, the mold temperature will take a couple of hours to stabilize.

Figure 7.31 shows that some dimensions may be more sensitive than others. Dimension B is not influenced by the changes in cooling time. However, Dimension A changes with the cooling time. The target value for dimension A is 2.8656". Therefore, the cooling time can either be set at approx. 17 seconds, or steel changes have to be made to run the mold faster and achieve the same dimensions. Identifying the lower and upper limits in Fig. 7.31 will also present a graphical representation of where the cooling time can be set.

Table 7.6 Cooling Study Worksheet

Cooling time (s)	Dimension 1 (in)	Dimension 2 (in)
8	1.552	0.253
10	1.553	0.253
12	1.554	0.254
14	1.554	0.254
16	1.555	0.254

Cycle time is the most important factor because it directly impacts the bottom line profit of a molding operation. In most cases, if the process is capable at shorter cooling times, changes in the mold steel may achieve the same dimensions at shorter cycle times. The cooling time is usually the major part of the molding cycle. Therefore, optimization of cooling time is critical for profitability. Figure 7.32 shows the typical cycle time break-down for all phases of the molding process. Part design and mold design can have a significant impact on the actual cycle time required for each phase shown in the pie chart of Fig. 7.32. A long thin part may only need a short cooling time; however, the mold may need to stay closed much longer to complete the screw recovery phase. Ideally, the recovery time would not exceed the time required to cool the part for ejection from the mold. However, part design, mold design, and machine selection can all have significant influence on the actual cycle times required for each phase of the injection process.

Table 7.6 is an example of a worksheet that can be used to document a cooling time study.

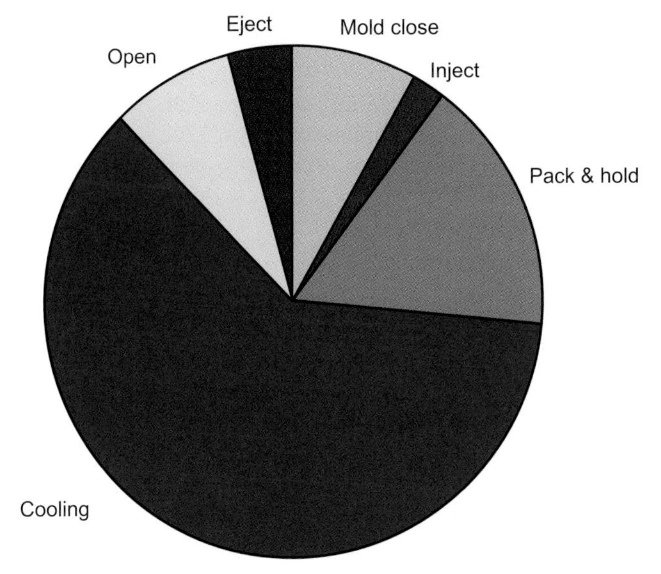

Figure 7.32 Typical cycle time break-down

7.6.7 Optimization of Screw Speed and Back Pressure

There are no scientific experiments that can be easily performed to optimize screw speed or back pressure. The screw speed should be set such that the screw always recovers before the end of the cooling time. For crystalline plastics, higher screw speeds are recommended because the melting of the crystallites requires high energy that can be supplied by the shear of the rotating screw. The heat from the heater bands may not always be sufficient to produce a homogeneous and uniform melt. Highly fiber filled plastics or those filled with long fibers typically require slower screw speeds to avoid fiber breakage. Some shear sensitive materials, such as PVC, also require low screw speeds. The optimum back pressure is the lowest pressure possible to keep the screw recovery time consistent and avoid any surface and/or internal defects visible on the part. Surface defects would include splay and internal defects would include bubbles and voids.

7.6.8 Post-Mold Shrinkage Studies

When the parts are ejected from the mold, they are below the ejection temperature but are still warm. If the temperature of the part is above the glass transition temperature, the molecules have enough energy to move and settle down into their equilibrium positions. This movement results in additional shrinkage and therefore a dimensional change of the part. If the glass transition temperature (T_g) of the plastic is below room temperature, the phenomenon continues until equilibrium between ambient and T_g-temperature is reached, causing even more excessive shrinkage. Because this shrinkage takes place outside the mold and after molding, it is called post-mold shrinkage. Thermoplastic elastomers are common materials with low glass transition temperatures exhibiting high post-mold shrinkage. The rate of shrinkage is directly proportional to the temperature of the plastic and follows an exponential decline with time after ejection. The highest shrinkage is therefore seen as soon as the part is ejected from the mold, because at this point the temperature is the highest. As the part starts to cool down, the rate of shrinkage also decreases. In some cases, part dimensions can take up to a few days to stabilize.

Because almost all molded parts are used in assemblies after ejection, the fit between the parts is very important. Therefore, part dimensions must have stabilized before assembly. If they continue to shrink after assembly, stresses can easily be induced into the assembly and cause premature failure. Figure 7.33 shows a part that continued to shrink after assembly and therefore induced stress in the assembly.

There are a number of factors that may affect post-mold shrinkage:

Glass Transition Temperature: The lower the T_g, the higher is the post-mold shrinkage. In plastics with T_g's above room temperature, post-mold shrinkage is low because there is not sufficient energy for molecular movement. High glass transition temperature plastics, such as PEEK, show almost no post-mold shrinkage.

Fillers and Additives: In the case of filled plastics, the presence of fillers prevents the shrinkage. Therefore, resins with a higher filler content exhibit lower post-mold shrinkage.

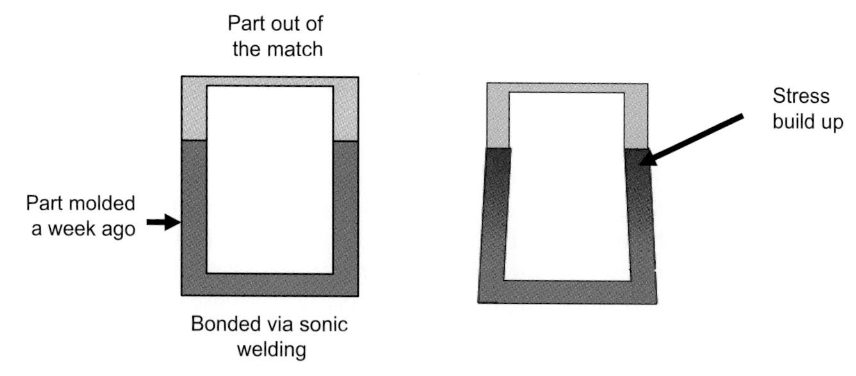

Figure 7.33 Post-mold shrinkage after assembly

Part Thickness: Part design also plays an important role with respect to the amount of post-mold shrinkage. Thicker part sections retain more heat and therefore can cause additional post-mold shrinkage. This can also lead to warpage, which is the result of uneven cooling in the part. If one area of the part has cooled down, while a thicker section of the same part is still cooling and undergoing some shrinkage, the thicker section will influence the thinner section and pull the part in the direction of the shrinkage, thus deforming the part.

Mold Temperature: The temperature of the mold provides the energy required to help bring the molecules to their equilibrium state. In case of crystalline materials, the mold temperature has to be maintained high enough to crystallize the plastic. If the mold temperature is not maintained at a sufficiently high level, the molecules freeze in place without reaching their equilibrium position. When these parts are ejected and are then subjected to higher temperatures during their service life, the molecules acquire the required energy and begin to find their equilibrium positions, thus causing shrinkage. This phenomenon can also occur when molded parts are stored prior to use. If mold temperatures were set lower than room temperature, for example 5 °C when the parts were produced and then stored at ambient warehouse temperatures of 30 to 40 °C, post-molding shrinkage can occur. When these parts are eventually removed from storage, they could exhibit some amount of warpage that did not exist at the time they were packaged.

Processing Conditions: Non-optimized process conditions can also induce stress in the part and cause post-mold shrinkage and warp. Under-packed parts commonly exhibit post-mold shrinkage while over-packed parts exhibit built-in stresses. These stresses are released after the part is ejected.

Annealing: Parts that have not reached their final state of equilibrium exhibit built-in stresses. The process of annealing can be considered as a forced post-mold shrinkage operation. In annealing, the parts are intentionally taken to higher temperatures to force the movement of the molecules out of their non-equilibrium positions and into their final equilibrium positions. In doing so all stresses are relieved. Annealing will result in some amount of shrinkage, depending on the factors described earlier.

A typical post mold shrinkage graph is shown in Fig. 7.34. The shrinkage exhibits an exponential drop: it is higher when the part is first ejected from the mold but then stabilizes over time.

Procedure to Measure Shrinkage

- Start molding with the established process parameters and let the process stabilize. The stabilization of the actual mold temperatures is an indication that the process has stabilized.
- Collect at least three shots as they are ejected from the mold and record the time each was ejected from the mold.
- After 15 minutes measure the parts from each shot and record the dimensions.
- Repeat after 30 minutes, 45 minutes, 1 hour, 2 hours, 4 hours, 8 hours, 24 hours and so on until a stable reading is achieved. The time interval between measurements can

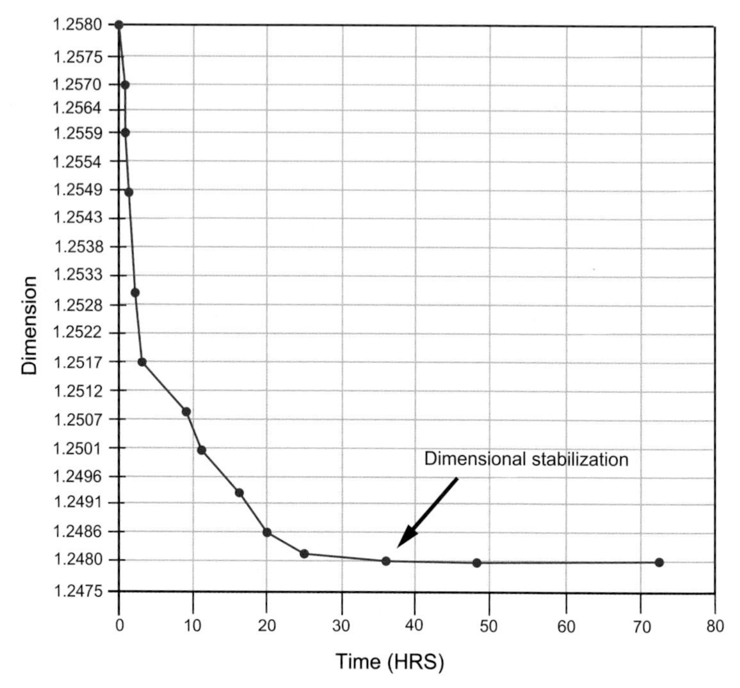

Figure 7.34 Post-mold shrinkage graph

increase as time after molding is getting longer. The goal is to collect the most measurements while the part is experiencing the highest rate of shrinkage, which is the time frame shortly after molding. As the part ages, the rate of shrinkage decreases and the time interval between measurements can be increased.

- Generate a graph of dimension versus time.

If there are multiple cavities, it may be difficult to measure all parts in a short time during the initial part of the experiment. In this case, measure any one cavity. If the cavity fill is balanced and the mold cooling for each cavity is the same, one can assume that the parts will behave in a similar fashion. The parts from the other cavities must still be collected with measurements being made at longer time intervals. Dimensional stability after longer relaxation times, e. g., at 2 hours or longer, is sufficient to measure the part dimensions.

How to Use this Information

Once the graph is generated, the time interval required for the part to become dimensionally stable will be obvious. Any secondary operations or direct part use must be done after the stabilization period. The stabilization time in Fig. 7.34 is approx. 36 hours.

7.7 Recommended Mold Function Qualification Procedure

The above sections described the science and the procedures to establish all process parameters. During each of the above steps, the most robust areas of each process parameter must be recorded. These parameter settings will contribute to the make-up of the final process. Once the process has been established, parts must be molded using these process settings. The machine must be set to the established process and a sufficient amount of parts must be run to check the process robustness and the part quality. There will always be some natural variation in the quality of the parts and therefore a statistically acceptable number of parts (usually 30) must be measured. Knowledge of this variation is essential to establish a capable process and consistent production.

Each procedure taken along the way may reveal an issue with the mold. For example, having a defect such as a short shot and flash at the same time is a clear indication of a mold problem. In this case, there is no process window and therefore it will be impossible to make an acceptable part. The mold must therefore be pulled out of the machine and fixed. This becomes a required mold trial iteration necessary for successful molding and production.

Likewise, other steps may reveal other issues, such as inadequate venting or improperly sized gates among others. An effort must be made to try and identify all issues with the fewest number of trials possible.

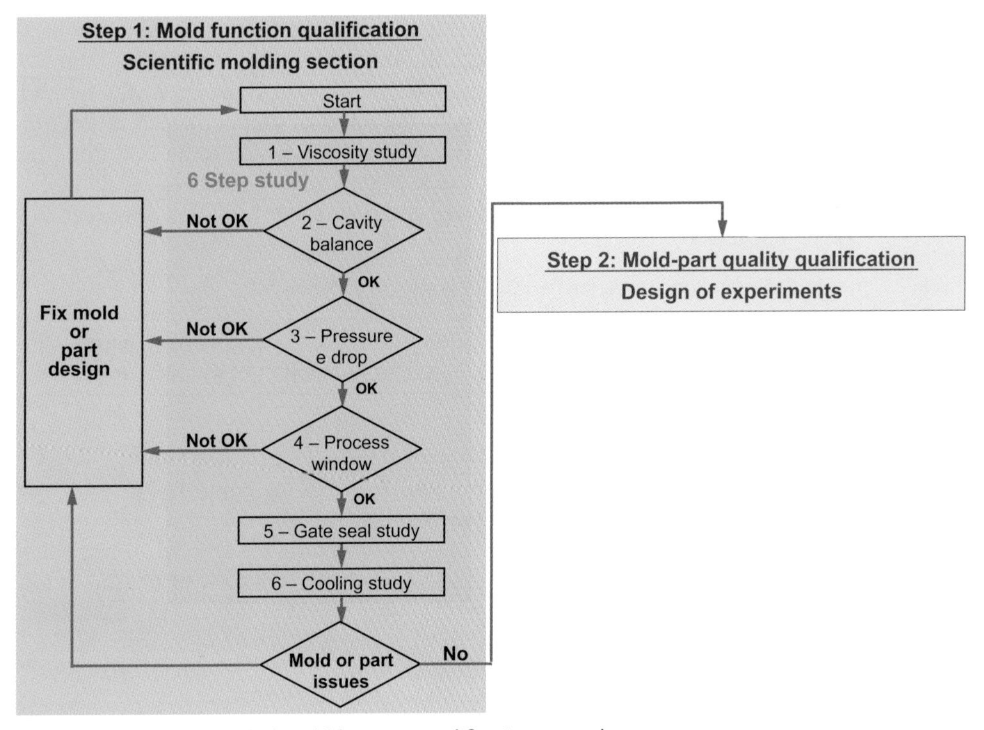

Figure 7.35 Recommended mold function qualification procedure

The procedure described is used to optimize the process and to make sure that the mold function is acceptable. It is therefore also called "mold function qualification procedure". Part dimensions were not the focus of this qualification, which does not mean that the dimensions should not be measured along the way. The dimensions must be measured and compared to the part drawing. If all the dimensions are acceptable and well within the quality requirements, the molding process can be recorded as final. This is the best case scenario. However, if any of the dimensions are not acceptable or the process is not deemed stable because of a large variation, the mold steel must be altered to center the dimension. Changing the process in turn could cause a process parameter to be in a non-robust area, causing inconsistencies. If the process window is wide enough, it is acceptable to move the process within the window. The flow chart in Fig. 7.35 shows the recommended mold function qualification procedure.

7.8 Recommended Adjustments to Maintain Process Consistency and Robustness

Once the process has been established, it is time to consider the dimensions of the parts. Collection of the quality data at the established process is crucial; therefore a good, statistically valid sample of parts must be measured. The minimum number of acceptable parts considered to provide reliable statistical data is 30. The data must be analyzed and a decision must then be made on how best to achieve the required quality standards. Table 7.7 describes the various scenarios and the recommended actions. In some cases, the natural variation can be larger than the difference between the lower and upper specification limits. For example, consider a part with the required nominal dimension to be 2.50 mm, the upper tolerance of 0.05 mm and the lower tolerance of 0.05 mm. The total available tolerance is therefore 0.10 mm. If 30 parts were measured and if the range (difference between the

Table 7.7 Recommendations for Centering Part Dimensions and Variations

Location of average – acceptance of variation	Adjust steel	and/or	Move nominal	and/or	Open tolerance
At nominal – acceptable	NA		NA		NA
At nominal – not acceptable	No		No		Yes
Closer to the spec limits – acceptable	Will help	and/or	Will help		No
Closer to the spec limits – not acceptable	Will help	and/or	Will help	and	Yes
Out of spec limits – acceptable	Yes	and/or	Yes		No
Out of spec limits – not acceptable	Yes	and	Yes	and	Yes

maximum and the minimum) was 0.12 mm, it will be impossible to mold all the parts within the required specifications, even if the average tolerance of the 30 parts was 2.50 mm. There will always be parts molded outside of specifications. In this case, opening up of the tolerances must be considered. Making any sort of steel adjustments will not help. If the range is less than the difference between the limits and the parts are out of specifications, then adjusting the mold steel will help to center the process and mold all parts within specifications. If the parts are functionally acceptable, the other option is to move the nominal dimension on the part print and leave the mold steel alone. This is a good option when the mold is not steel safe and the decision should be made by the part designer.

7.9 Process Documentation

Injection molding is primarily a heat transfer process, involving variables related to speed, pressure, time, and temperature. It is important to record each of these process parameters in order to duplicate the process during the subsequent trials and final production. The mold qualification worksheets, machine setup sheet, water line diagrams, mold temperature maps, setup instructions, and operator instructions must all be documented and/or updated during each trial. A section in Chapter 10 describes this documentation.

Of particular relevance here is the mold qualification checklist. A master checklist must be developed for mold trials, depending on the requirements of every molding operation. During every mold trial, the checklist must be used to evaluate the mold and the process. This assures that every part of the mold or process is evaluated and all the problems and/or suggested improvements are documented. This documentation can then be passed on to the mold maker or those involved with project improvements. A sample mold qualification checklist is provided in the Appendix.

References

1. Kulkarni, S.M., Hart, David, *SPE ANTEC Tech Papers* (2003) p. 736
2. Mertes, S., Carlson, C., Bozzelli, J., Groleau, M., *SPE ANTEC Tech Papers* (1997)

Suggested Reading

1. Osswald, T.A., Turng, L., Gramann, P.J., *Injection Molding Handbook* (2007) Hanser, Munich
2. Beaumont, J.P., *Runner and Gating Design Handbook* (2007) Hanser, Munich
3. Beaumont, J.P., Nagel, R., Sherman, R., *Successful Injection Molding* (2002) Hanser, Munich
4. Rosato, D.V., Rosato, D.V., *Injection Molding Handbook* (2000) CBS, New Delhi, India
5. Kulkarni, S.M., *SPE ANTEC Tech Papers* (2003) p. 736
6. Cogswell, F., *Polymer Melt Rheology* (1981) John Wiley, USA
7. Dealy, J., Wissbun, K., *Melt Rheology and its Role in Plastic Processing Theory and Applications* (1990) Van Nostrand Reinhold

8 Design of Experiments for Injection Molding

Planned experiments have been around for centuries. In the 17th century a doctor planned some experiments to find a cure for an ailment of that day. He engaged various patients and various combinations of medicines to find the right cure. Planned experiments were used in agriculture to find the right combination of factors, such as soil type and fertilizer, to produce the highest yield of crops. Since the time involved in waiting for the results of these agricultural experiments was one complete season, which in some cases was one complete year, the technique of a planned experiment was very helpful. As planned experiments became popular as time saving and efficient techniques, people with a background in mathematics and statistics became involved and developed them further. Each came up with techniques that helped in analyzing different types of data in different scenarios. The people who have been most associated with the development of Design of Experiments are G. Taguchi and Ronald Fischer. The field of designed experimentation or DOE is vast and considered a specialty of its own. Within the context of this book we will focus on "Factorial Experiments". Factorial experiments work very well for injection molding. The use of any procedure (not just a DOE procedure) must be backed by a complete understanding of the underlying principles, which helps not only in the understanding of the analysis but also more importantly in the interpretation of the analysis. Factorial experiments, their analysis, and their interpretation are easy to comprehend and do not require a very strong mathematical background.

8.1 Parameters in Injection Molding

Applying and using DOE in injection molding is relatively simple compared to its use in other manufacturing or production processes, because here most responses to process changes are linear. For example, there is a direct relationship between part dimension and holding pressure. If the dimensions of a sample part molded at two holding pressures are known, it is safe to predict that the part dimension produced when molded at the average of these two pressures will lie at the average of the two dimensions, as shown in Fig. 8.1.

The same relationship exists between other responses and processing parameters. In injection molding, all plastic material variables are related to speed, pressure, time, and temperature. The linear response can be explained with the help of specific volume versus temperature graph discussed earlier. In Fig. 8.2 it is shown with the molding area corresponding to the injection, pack and hold phase of the molding cycle. In this area the curves are linear regardless whether the material is amorphous or crystalline.

A similar graph for a PBT-PC blend generated at different pressures is shown in Fig 8.3. Such a graph is also called a PVT graph.

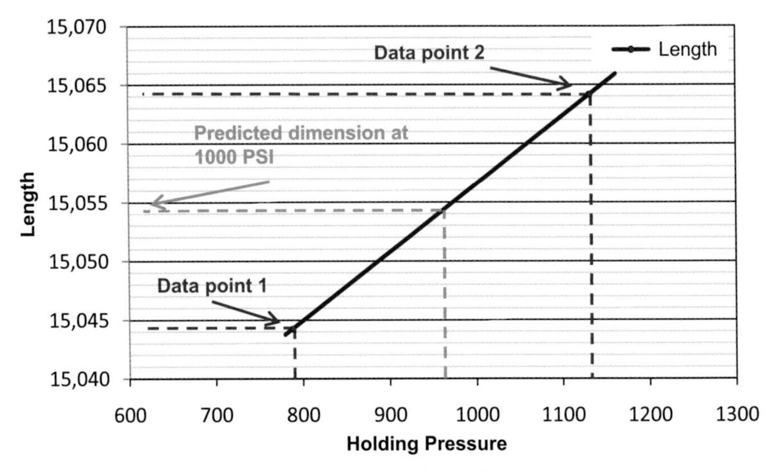

Figure 8.1 Relationship between holding pressure and part dimension

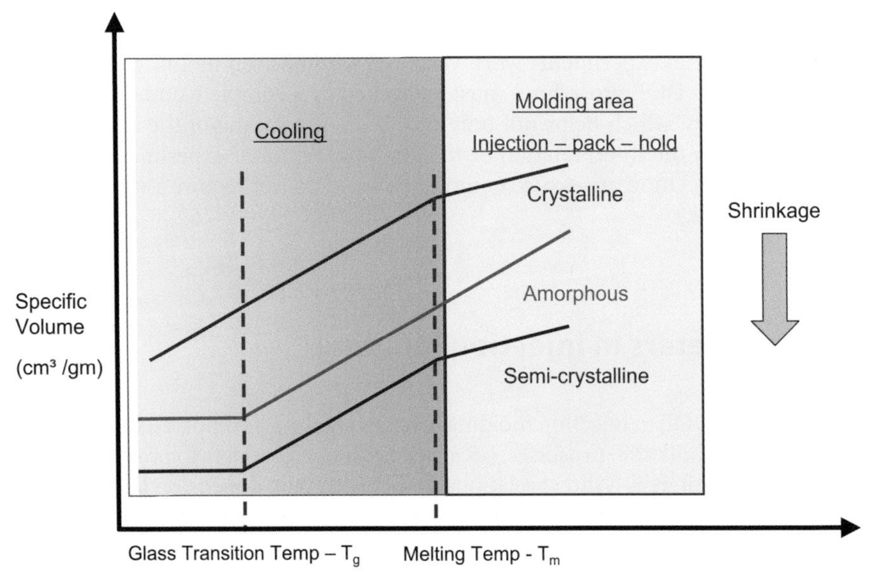

Figure 8.2 Specific volume versus temperature graph showing the area corresponding to the injection, pack and hold phase

Shrinkage is the change in volume as the plastic is being cooled. Based on the linear relationship between the volume and the temperature it is safe to assume that the part dimensions also respond linearly to process parameters such as temperatures and pressures. Faster mold fill rates will result in lower heat loss in the melt before it reaches the end of fill. Therefore, the plastic characteristics are in the top right hand side quadrant of the PVT graph, but are still in the linear area. Increasing or decreasing the fill speeds will result in a proportional change in dimension. During the cooling time, the melt is now out of the shaded molding

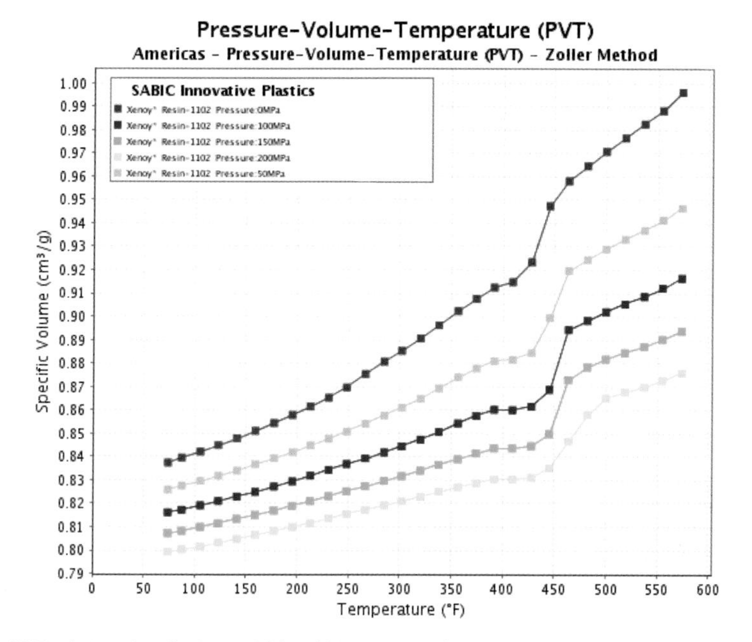

Figure 8.3 PVT relationship for PBT-PC blend (Courtesy: Sabic Innovative Plastics)

area shown in Fig 8.2. The melt begins to solidify and the volume begins to decrease following the plot shown in Fig 8.2. The plot is linear until the plastic reaches its glass transition temperature (T_g) where the curve exhibits an inflection point. If the part is ejected above its T_g, the relationship between the specific volume and the temperture is linear. If it is ejected below the T_g, there will be some non-linearity. For efficient injection molding, parts must always be ejected at a material-specific ejection temperature that is always above the T_g. If the parts are being ejected below the T_g, the molding is not efficient and additional time is added to the cycle unnecessarily. The parts must always be ejected above or close to the T_g, keeping the cooling curve in the linear region of the PVT graph. The plastic will continue to shrink and therefore a post-mold shrinkage study must be done on the parts. If the molding process was robust and consistent, the post-mold shrinkage will also be consistent, producing consistent parts. The application of the above concept is discussed in the following example explaining the concept of DOE.

A note on the proportional changes mentioned above is required here. These relationships can be directly or inversely proportional to each other. For example, packing pressure can increase the length of a part, but in some cases, such as an internal diameter of a part, there can be a decrease in the diameter of the part with increase in the packing pressure. Prediction is not easy and experimentation is required.

Design of Experiments: Definition

The simplest description of Design of Experiment (DOE) is a planned study. For example, studying the effect of holding pressure on the length of the part is a designed experiment. The length of the part at a low holding pressure and a high holding pressure is measured and is then plotted as a function of the holding pressure, see Fig 8.4. This is the most basic DOE that can be performed.

When considering two parameters, holding pressure and melt temperature, and their influence on the length, we need to perform four experiments and determine the length at the following holding pressure / melt temperature combinations: low – low, low – high, high – low, and high – high, see Fig. 8.5.

If we add one more parameter to this, for example mold temperature, we end up with eight necessary experiments. We are repeating the above four experiments at a low and high value of the mold temperature, resulting in eight experiments, as shown in Fig. 8.6. As the number of parameters to be studied increases, the number of experiments increases.

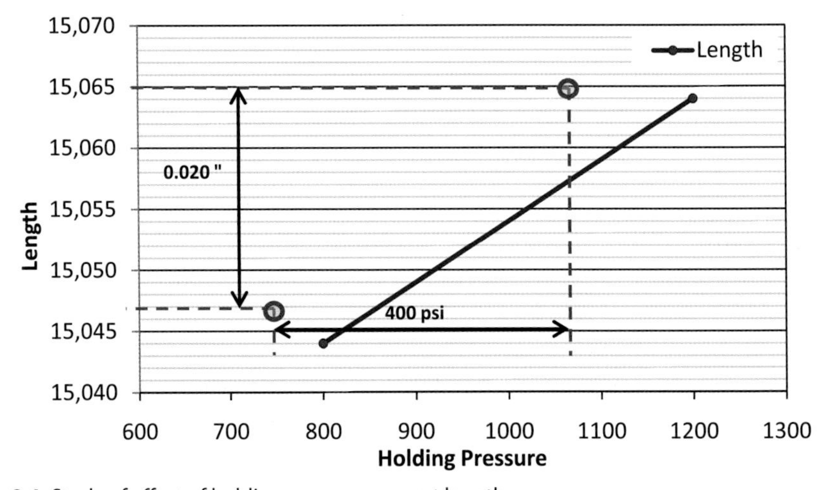

Figure 8.4 Study of effect of holding pressure on part length

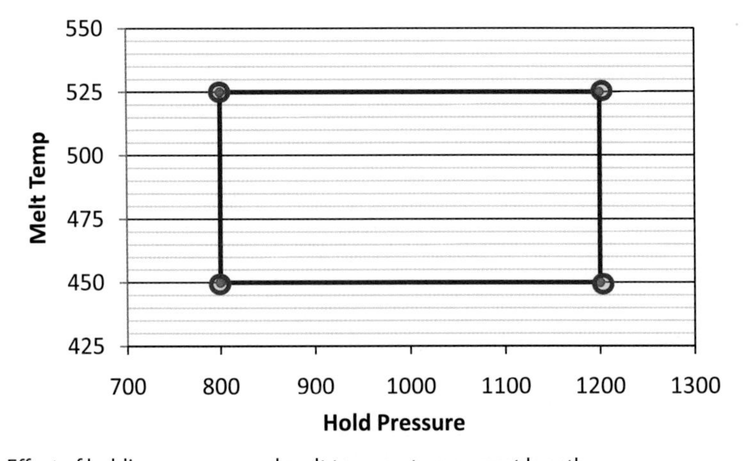

Figure 8.5 Effect of holding pressure and melt temperature on part length

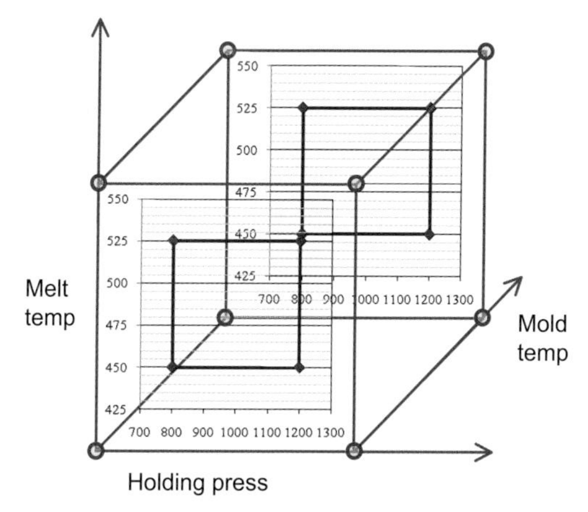

Figure 8.6 Effect of holding pressure, melt temperature, and mold temperature on part length

8.2 Terminology

8.2.1 Factor

Any input to the process is a factor. Therefore, all processing parameters that are input to the molding machines are factors. Example: Holding pressure and melt temperature. Factors can be set to a desired value on the machine controller or can be selected from available options, such as a particular lot of material. Factors are classified as follows:

Control Factors: Can be controlled and changed when required, e.g., mold temperature

Noise Factors: Cannot be controlled, e.g., lot-to-lot variation

Constant Factors: Are not changed during the study, e.g., back pressure

Quantitative Factor: Can be continuously changed in increments, e.g., holding pressure

Qualitative Factor: Can be changed in discrete levels, e.g., material lots

8.2.2 Response

Any output from a process is a response. A response is the result obtained at the various level settings to which the factors are set to during an experiment. The value or attribute of a response depends on the setting of a factor. The response cannot be controlled directly. To get a required value of a response, the value of the factor must be changed. Part dimensions, fill times, cavity pressure, or the amount of splay are all examples of responses. Responses are classified as follows:

Quantitative Response: Are represented by numbers, e.g., length, weight

Qualitative Response: These are also called attributes; they are not represented by numbers and describe the condition of the response, e.g., splay, color

8.2.3 Level

A level is the number of points selected for the factor studied. For example, if we choose a low and a high value of holding pressure, there are two levels. If we choose a low, center, and a high value, there are three levels. The number of levels is chosen based on the type of response to the factor. In injection molding, most responses to the factors are linear. This means that when conducting a study at a high level and a low level, the response at a medium value of these two factors can be predicted to be the average of the responses. If the part length at 500 psi of holding pressure is 1.10 inches and at 1500 psi it is 1.20 inches, we can conclude with a high level of confidence that the length at 1000 psi will be 1.15 inches. There are some exceptions, such as cooling time, where a particular dimension may plateau off. In some cases, the holding pressure may also plateau off at very high values. This is where a good amount of practical experience and engineering knowledge can help. In most cases, a two-level experiment followed by a confirmation study suffices most needs. Figure 8.7 shows some of the factors, levels and responses in injection molding.

8.2.4 Designed Experiment

A designed experiment is a study in which purposeful changes are made to the factors and the corresponding effects on the responses are recorded. The responses are analyzed and

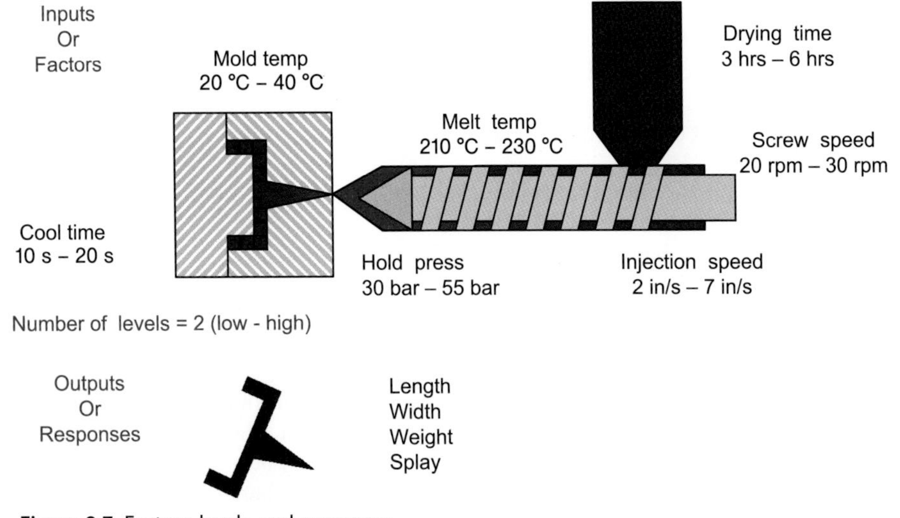

Figure 8.7 Factors, levels, and responses

the results are used to optimize the process to make it robust. The experiments mentioned earlier are all examples of designed experiments. Table 8.1 shows a matrix for a three-factor, two-level, and two-response designed experiment.

Table 8.1 Matrix for a 3-Factor, 2-Level, and 2-Response Experiment

Runs	Experimental run settings			Responses	
	Mold temp (°C)	Cooling time (s)	Holding press (bar)	Length (mm)	Diameter (mm)
1	40	30	30	144.73	6.35
2	40	30	55	144.40	6.15
3	40	20	30	144.60	6.32
4	40	20	55	144.30	6.15
5	20	30	30	144.83	6.37
6	20	30	55	144.50	6.16
7	20	20	30	144.65	6.32
8	20	20	55	144.34	6.16

8.3 Relationships Between the Number of Factors, Levels, and Experiments

The main goal of performing various experiments is to understand the effect of the factors on the final quality of the part. Therefore, when the number of factors increases, the number of experiments also increase, see Table 8.2.

Table 8.2 Number of Experiments Based on the Number of Levels and Factors

Levels	Factors								
	1	2	3	4	5	6	7	8	9
2	2	4	8	16	32	64	128	256	512
3	3	9	27	81	243	729	2187	6561	19683

With two factors at two levels, the number of required experiments is 4. The relationship between the number of experiments, factors and levels is given by the following equation:

$$n = l^f \tag{8.1}$$

where,

n = Number of experiments

l = Number of levels

f = Number of factors

For example, for 3 levels and 4 factors, we have 3 × 3 × 3 × 3 = 81 experiments. Running a large number of experiments can get expensive and very time consuming. However, inspection of all parts and collecting the data for the number of dimensions under consideration would be even more time consuming. In injection molding, it is common to work with multiple cavities and collect statistical data for several part dimensions. If this information needs to be collected, the required time and the effort involved can increase significantly. There are multiple designs of experiments available where the number of experiments can be reduced, yet reliable data can be obtained from them. These experiments are subsets of the maximum number of required experiments mentioned earlier and this is where the advanced techniques of designed experiments become a powerful tool.

Over the years, a number of researchers presented various ways to design fewer experiments, analyze the data, and provide reliable results. Each have their own advantages and disadvantages and are suitable to particular types of experiments. Some of the popular designs in injection molding are screening experiments, Taguchi designs, Plackett-Burman designs, and Box-Behnken designs. Although screening experiments are usually followed by full factorial experiments, in injection molding that is usually not necessary because there are only a handful of factors that truly make a difference in the quality of the part. Factorial experiments and the Taguchi methods are most commonly used. Analysis of the results does not require a lot of mathematical and/or technical background, which makes the Taguchi screening methodology very user-friendly. Once understood, common spreadsheets programs, such as MS Excel, can be used to analyze the data. However, because of the time involved in creating such a worksheet, it may be best to use software specially created for such purposes. Some common software programs for DOEs are mentioned in the bibliography.

To understand DOEs further, we will introduce a number of concepts in the following.

8.4 Balanced Arrays

Table 8.3 represents a 2-factor experiment with 2 levels for each factor. Therefore, an experiment resulting in 4 test runs is required.

An array is said to be orthogonal if it meets the following two conditions:

a) *For each factor, an equal number of highs and lows are tested.*

In this case, there are two factors, holding pressure and melt temperature.

Hold pressure – 2 highs and 2 lows are tested
Melt temperature – 2 highs and 2 lows are tested

b) *For each level within a factor, equal number of highs and lows for every other factor are tested.*

In this case:

For low hold pressure – 1 low of melt temperature and 1 high of melt temperature
For high hold pressure – 1 low of melt temperature and 1 high of melt temperature
For low melt temperature – 1 low of hold pressure and 1 high of hold pressure
For high melt temperature – 1 low of hold pressure and 1 high of hold pressure

Table 8.3 Two-Factor, Two-Level Experiment

Experiment No	Hold pressure	Melt temperature
1	Low	Low
2	High	Low
3	Low	High
4	High	High

The matrix shown in Table 8.3 is an orthogonal array because it meets both conditions. There is a balance between the number of highs and lows of each factor of the experiment. If we consider the same concept of orthogonal arrays from a mathematical aspect, replacing the highs and lows with numerical values, low would be replaced with –1 and high would be replaced with +1. replacing the names of the factors and calling them A and B will result in Table 8.4.

Now, we will perform the following operations on the entries in this table:

Table 8.4 Two-Factor, Two-Level Experiment

Experiment No	A	B
1	–1	–1
2	+1	–1
3	–1	+1
4	+1	+1

Table 8.5 Orthogonal Arrays

Experiment No	A	B	A × B
1	–1	–1	+1
2	+1	–1	–1
3	–1	+1	–1
4	+1	+1	+1
Sum	**0**	**0**	**0**

1. Create a new column called AB whose cells are the result of the product of the cells of A and B.
2. Add the cells in each of the columns in a new row.

In Table 8.5, the sum of each of the columns equals zero. Such an array is called an orthogonal array. Mathematically, the array is said to be vertically and horizontally balanced, if the sum of each of the columns is zero. This means that there are an equal number of highs and lows, making it a balanced column. When each column is balanced, the experiment is also considered balanced.

The importance of orthogonal arrays lies in the fact that each column and row can give us a unique set of reliable information and this can be used as an advantage to reduce the number of runs in our experiments. We will discuss this fact in more detail later on in this chapter.

8.5 Interactions

We will use the following example to explain what interactions within the concept of Design of Experiments represent: as the humidity and temperature increase, human comfort level decreases. For this experiment, the comfort level is graded on a scale from 0 to 10, where 10 is the most comfortable. When the humidity is low, a change in the temperature from 20 °C to 28 °C changes the comfort level from 9 to 8. In fact, most people do not even notice the change. However, when the humidity is high, the comfort level even at 20 °C is relatively low and with the same change in the temperature to 28 °C the comfort level falls to 2. People in coastal towns often face such discomfort, because of the high humidity even at low temperatures. If we plot these results, as shown in Fig. 8.8, it is clear that the change in comfort between two temperatures at low humidity is different from the change in comfort at high humidity. There is a drop of 1 point at low humidity and a drop of 4 points at high humidity. Therefore, the amount of change in comfort due to temperature is dependent on another factor, the humidity. In technical terms this means that there is an interaction between temperature and humidity when it comes to human comfort levels.

Expt No	Humidity %	Temp	Comfort Level
1	20	20	9
2	20	28	8
3	80	20	6
4	80	28	2

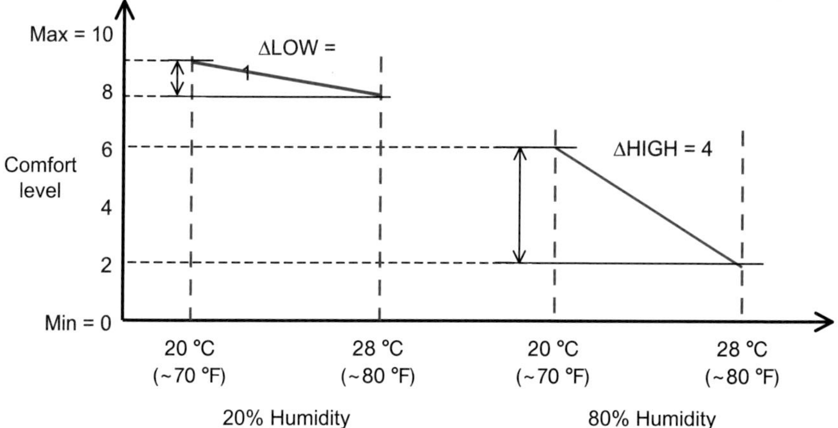

Figure 8.8 Interaction between humidity and ambient temperature regarding human comfort level

An example for no interaction between temperature and humidity is the pressure inside a container or an automotive tire. The change in tire pressure with an increase in temperature from 20 °C to 28 °C will be the same regardless whether the humidity is 20% or 80%. Interactions can be non-existent, mild, or strong as shown in Fig. 8.9.

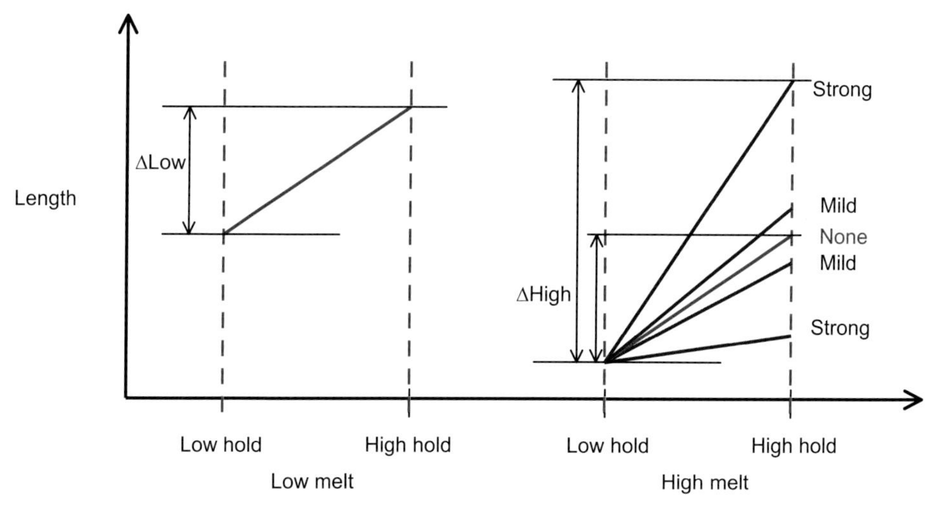

ΔLow differs considerably from ΔHigh: Strong Interaction between the factors
ΔLow differs a little from ΔHigh: Some Interaction but not strong

Figure 8.9 Types of interactions

Table 8.7 Introducing Confounded Factors in the Interaction Columns

	Non-confounded factors			Confounded factors			
Experiment No	**Holding pressure**	**Melt temperature**	**Mold temperature**	**Cooling time**	**Injection speed**	**Holding time**	**Screw speed**
1	–1	–1	–1	–1	+1	+1	+1
2	+1	–1	–1	+1	–1	–1	+1
3	–1	+1	–1	+1	–1	+1	–1
4	+1	+1	–1	–1	+1	–1	–1
5	–1	–1	+1	+1	+1	–1	–1
6	+1	–1	+1	–1	–1	+1	–1
7	–1	+1	+1	–1	–1	–1	+1
8	+1	+1	+1	+1	+1	+1	+1
Sum	**0**	**0**	**0**	**0**	**0**	**0**	**0**

This shows that the biggest advantage of confounding is in the reduction in the number of experiments required to study the process. Based on the equation

$$n = l^f,$$

a 7-factor, 2-level study would need 128 unique experiments, but with the process of confounding, these can be reduced to a mere 8 experiments. The general assumption when using confounding is that the interactions are minimum or non-existent.

8.7 Randomization

The experiments in the set above are all arranged in a particular order. For example in Table 8.7, there are eight experiments, of which the first four are done with a low mold temperature and the next four are done with the high mold temperature. The rest of the factors also have a certain regular order. Randomization is not following any such order and completely selecting the experimental order at random. Each row is an experiment and randomization means running the experiments in a random order. Some experts advocate randomization in order to exclude any external effects that cannot be controlled. For example, if the first four of the above experiments are done in the morning, when the ambient temperature is low, and the remaining eight are done in the afternoon, when it is hotter, the ambient temperature can have an effect on the response. The results will be compounded by the effect of the temperatures in the morning and the afternoon. This will make it difficult to separate the effect from morning and afternoon temperature change from the effect of the mold temperatures that were also different in the morning and the evening. There could be other factors related to ambient temperature that one may not be aware of. Tower water temperature, operator skills (not necessarily time efficiency), and material lot variation are examples of these factors. Therefore, mixing up or 'randomizing' the experiments will help to even out some of these effects, although not systematically. Randomization also helps to evaluate the robustness of a setting. For example, consider a knob with graduations being used to set the holding pressure and a setting of 6 on the knob yielded a certain value of holding pressure. Setting it once in the beginning of the experiment and leaving it for a set of experiments will deliver the same consistent pressure. However, changing it to another value and then going back to 6 may yield another value of hold pressure. This demonstrates the robustness of the setting and going back and forth with the settings helps the evaluation of the equipment settings.

With a large number of experiments in injection molding, it will be a good idea to randomize at least part of the experiments. For short runs, randomization is not required. It is always good to make a note of the other factors that are constant. Noting the operator's name, material lot number, etc. are part of good documentation procedures.

8.8 Factorial Experiments

The experiments and tables used in the earlier discussion are all factorial experiments. The design with the maximum number of possible experiments without confounding is called a full factorial experiment. For example, a 2-level, 4-factor design would require 16 experiments in a full factorial design. Partial factorial experiments are designs that have fewer experiments using the technique of confounding. Reliable data can be obtained from partial factorial experiments to analyze the effect of the factors.

8.9 Data Analysis

Next, basic analysis with the available data will be discussed. Today, most analysis of DOE data is performed by computer programs that generate all the information and graphs in a matter of seconds. A typical analysis provides the following information:

- Factors that most influence the quality of the part
- The robustness of the quality of the part
- Prediction of the most optimized process
- Prediction of the capability of the process within the range of the parameters studied

Table 8.8 Experimental Settings and Response Data for a 3-Factor, Full Factorial Experiment

Runs	Experimental run settings			Responses	
	Mold temp (°C)	Cooling time (s)	Holding press (bar)	Length (mm)	Diameter (mm)
1	40	30	30	144.73	6.35
2	40	30	55	144.40	6.15
3	40	20	30	144.60	6.32
4	40	20	55	144.30	6.15
5	20	30	30	144.83	6.37
6	20	30	55	144.50	6.16
7	20	20	30	144.65	6.32
8	20	20	55	144.34	6.16

To explain the results and the analysis it is best to consider an experiment. Consider the DOE results shown for the experiment in Table 8.8.

The three selected factors were mold temperature, cooling time, and holding pressure. There were two levels for each factor. The response was the length of the part and an internal diameter. Based on engineering knowledge and past experience, the order of importance for the factors was holding pressure – cooling time – mold temperature. However, if these three factors are arranged in the order of 'difficulty of change,' then they are rearranged in the order mold temperature – cooling time – holding pressure. 'Difficulty of change' simply means how soon the change is seen in the process. For example, once the mold temperature is changed, depending on the size of the mold, it may take anywhere from 15 minutes to an hour to see a change in the actual mold temperature. So this is a 'difficult' change. On the other hand, holding pressure is an easy change, because once the setting is changed, the change will be reflected in the next cycle. If the experiments are not going to be randomized, it is a good idea to have the order of the factors in the order of difficulty. This is an efficient way of getting all the experiments done by making the least amount of difficult changes during the experiments. So, in the above array, we will have to change the mold temperature (difficult change) only once compared to the holding pressure (easy change), which will have to be changed seven times after the first experiment.

With the help of this matrix, the experiments are performed. It is best to collect as many data as possible for accurate analysis. Typically, if statistical analysis is to be done, at least 30 parts from each cavity must be checked and the data recorded. However, as the number of cavities and recorded dimensions for a part increase, the amount of work to collect the data increases and this can sometimes become prohibitive. Performing full factorial experiments with a larger number of factors will again increase the number of experiments and therefore the number of measurements. In the experiment under consideration, we are going to measure the length of the part. We will take an average of 5 samples for each experiment.

The results are typically displayed in one of the following forms:

8.9.1 Tornado Charts

Tornado charts are similar to the commonly used Pareto charts. In DOE, a tornado chart is a bar chart in which the effect of each factor is plotted in descending order of magnitude. A factor that has a direct effect is displayed on the positive side of the y-axis, while a factor that has an inverse effect on the response is plotted on the negative side of the y-axis. A tornado chart is one of the most important and informative charts in DOE. It will display the factors that have the most effect on the quality of the part in descending order of importance. For the experiment under consideration, the tornado chart is shown in Fig. 8.10.

Figure 8.10 shows that the factor with the most impact on the length of the part is the holding pressure. For the levels of the holding pressures tested (between 30 and 55 bar), the average change in length was 0.3175 mm, compared to the 0.1425 mm change observed between the cooling time levels of 20 and 30 seconds. Change in mold temperature had the least effect on the length of the part. The average change in length between 20 °C and 40 °C was only 0.0725 mm. Notice that the mold temperature bar is on the negative side of the axis. This means that when the mold temperature increases, the part length decreases. Positive values, as seen for holding pressure and cooling time, indicate that as these factors are increased, the length of the part also increases. The interactions are also plotted and are very insignificant. In Figure 8.10, holding pressure and cooling time have a significant effect and we can consider that the effect of the other factors is insignificant. The tornado chart for the diameter is shown in Fig. 8.11.

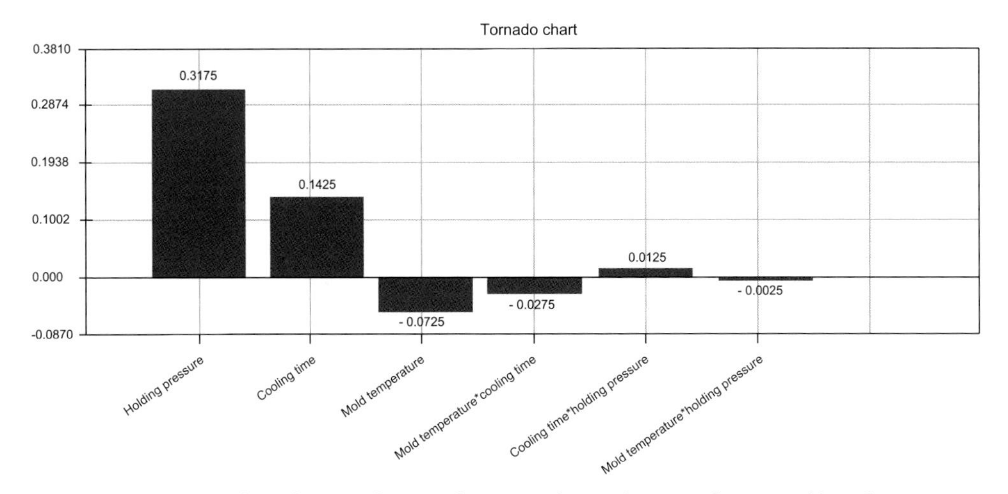

Figure 8.10 Tornado chart showing direct and inverse relations between factors and length response

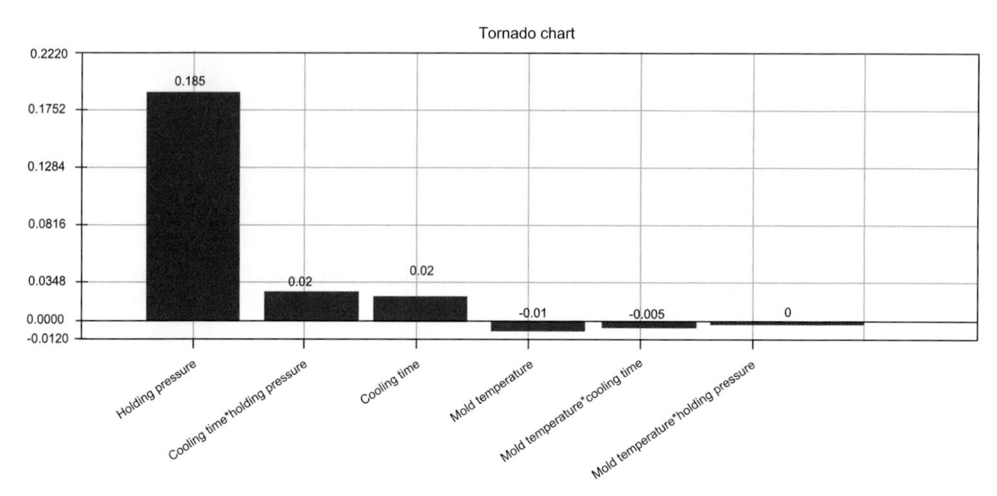

Figure 8.11 Tornado chart for diameter of the part

8.9.2 Contour Plots

Contour plots display contours that represent a constant value of a given response on a graph of two of the selected factors on the x and y axis. On any given point on a given contour, the value of the response will be constant, regardless of the values on the axes. Figure 8.12 shows a contour plot based on holding pressure and cooling as the factors.

Consider the highlighted contour that has a value of 144.65, which is also the nominal value. Anywhere on this contour, the value will always be 144.65. A combination of 29 s of cooling time and 42.5 bar of holding pressure (Point A) will produce a part with the same dimension as the combination 22 s and 52 bar (Point B). When the nominal value and the specifications are plotted on the contour plot, the process window inside of which dimensionally acceptable parts can be molded is visible. In Fig. 8.12, the orange contours represent the upper and lower specifications. Contour plots help to determine the extent of dimensional process windows. The different types of process windows are explained in Chapter 9.

8.9.3 Prediction Equation

A relation between the response and the factors can be established by mathematical means. Figure 8.1 showed the increase in the length of a part that resulted due to an increase in the holding pressure. If we assume the relationship to be linear, we can generate an equation in the form $y = mx + c$, where y is the length and x is the holding pressure. The value of m and c can be determined with the help of the two points. Finding the coefficients and constants of this equation, the dimension at any value of holding pressure or the holding pressure required for a desired dimension can be predicted, see Fig. 8.13. This is the basis of prediction equations that involve all factors and their interactions. Prediction equations are beneficial in selecting process parameters for robust processes and for hitting target dimensions. It is not

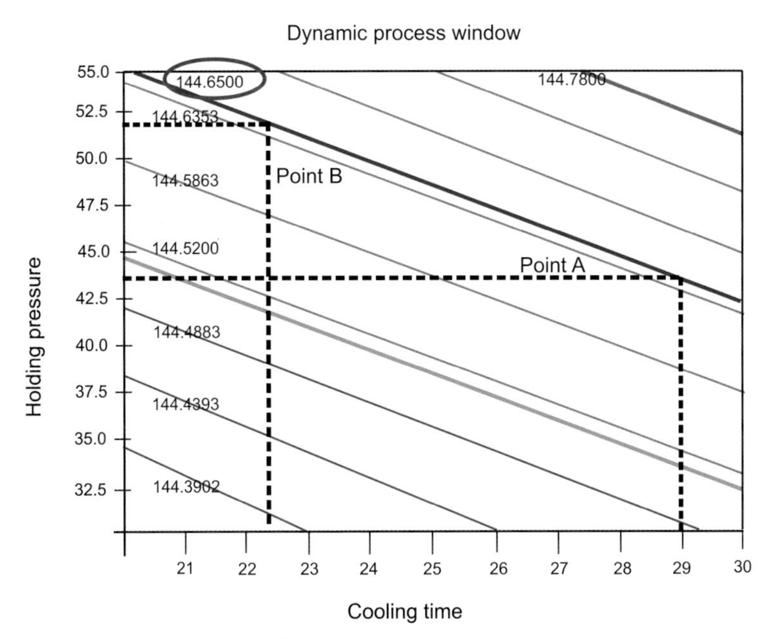

Figure 8.12 Contour plot for length of the part

always possible to hit target dimensions on all the responses. While one dimension is in tolerance, another may be out of tolerance. Overlaying contour plots or looking at composite and desirability functions is a good way to estimate the robustness and the process capabilities.

Mold Temperature	Cooling Time	Holding Pressure
30	25	42.5

Response Name	+ TOL	NOM	- TOL	Predicted Value
Length	0.13	144.65	0.13	144.544
Diameter	0.2	6.25	0.2	6.25

Mold Temperature	Cooling Time	Holding Pressure
30	25	30

Response Name	+ TOL	NOM	- TOL	Predicted Value
Length	0.13	144.65	0.13	144.385
Diameter	0.2	6.25	0.2	6.16

Figure 8.13 Prediction equation for length and diameter

8.9.4 Process Sensitivity Charts

Process sensitivity charts are line graphs that provide a quick glance of the position of the response for each experiment. Such a chart can be considered as a visual of the sensitivity of the response to a change in factors for each experiment. The upper specification limit (USL) and the lower dimensional specification limit (LSL) as well as the nominal are shown on the graph. If for all experiments the response is within the specification limits, the response is considered stable and not affected by the process. Such a dimension is shown in Fig. 8.14. If this was considered a critical dimension to be checked regularly during production, a case can be made to eliminate this dimension for in-process inspection and it could instead be checked at startup only. Figure 8.15 shows a dimension that is easily affected by process changes. Composite process sensitivity charts display all responses on one screen to give a snap shot of all dimensions at one time.

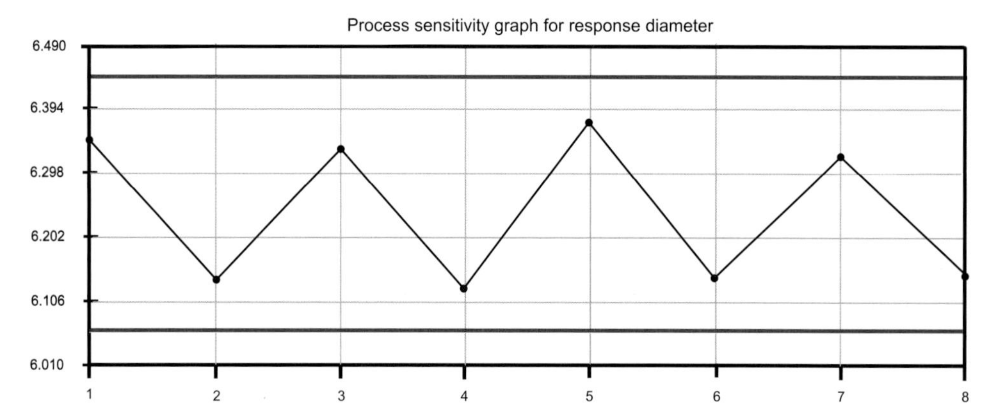

Figure 8.14 Process sensitivity chart for the diameter

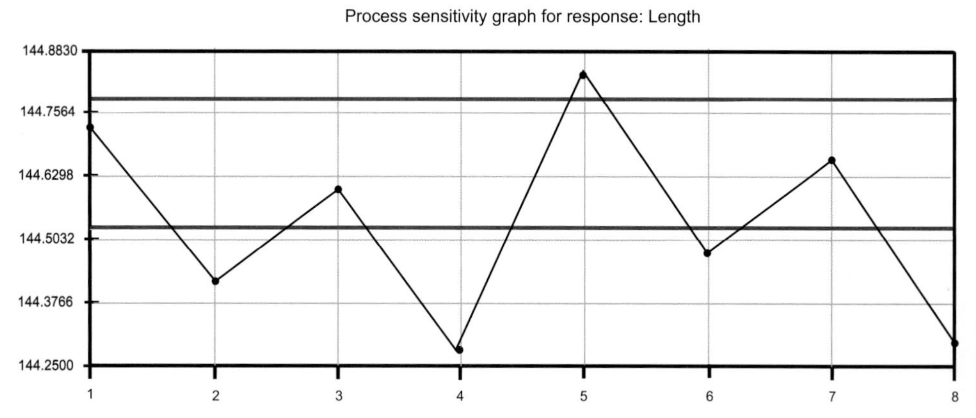

Figure 8.15 Process sensitivity chart for the length

8.10 Using the Results from DOE

The graphs and results from the above charts and equations can be used in a number of ways.

8.10.1 Process Selection

The contour plots will display the constant response curves together with the specification limits. The goal should be to achieve the nominal value of the response and be in a robust processing area, see Fig. 8.12, where the LSL, USL, and the nominal are shown. To achieve the nominal, various combinations of the holding pressure and cooling time exist; for example, at 29 seconds and 42.5 bar or at 22 seconds and 52 bar. Any one of these combinations can be selected as process parameters because both are in a robust molding area. In injection molding, lower cycle times increase production efficiency. Therefore the lower cooling time must be preferred and selected.

8.10.2 Cavity Steel Adjustment

If the nominal values and the specifications are located towards the corners or boundaries of a contour plot, it may be difficult to sustain production because the process may not be robust enough to mold the parts within specifications. For the part shown in Fig. 8.16, the processing window between a short shot and out-of-specification is only about 50 psi (3.44 bar). The process must be kept within this 50 psi window to avoid shorts and parts out of specification. The process window is small and therefore not robust.

If the process variation is large, it is easy to produce parts that are either short or out of specifications. In such cases, without taking the dimensions into consideration, a process with a sufficient process window must be selected. Once it is determined that the process can produce parts with the required dimensional consistency (not the actual dimension), the steel

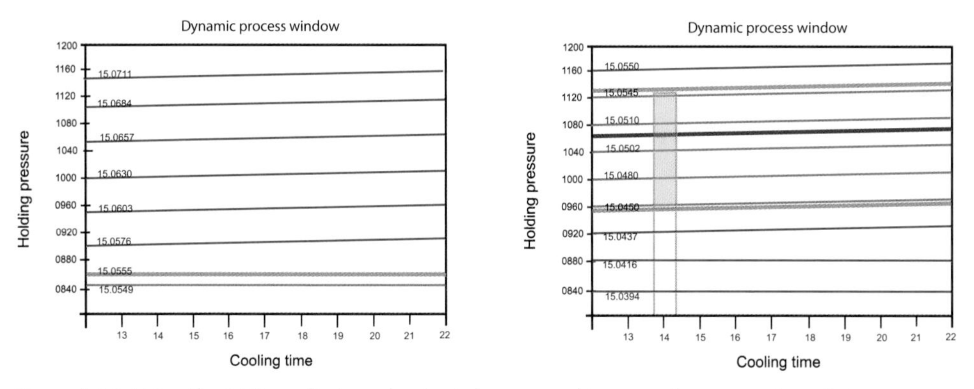

Figure 8.16 Using the DOE results to make steel changes and increase the processing window

must be adjusted to bring the parts within the required dimensions. For example, for the part in Fig. 8.16, there is a danger of short shots because the holding pressure must be kept low to achieve the required dimension. If the pressure is increased, the parts will be out of specifications but the chances of short shots will be greatly reduced. Running at the center of the process window will produce parts that have the least or zero possibility of short shots. Parts must be molded at the center of the window and the dimensions measured. Depending on the dimensions, the mold steel must now be changed to bring the molded parts within the required tolerance. This will result in a robust process that will produce parts within the desired specifications.

8.10.3 Process Adjustment Tool

The factors used in the DOE were selected based on the assumption that they were the most influential to the part quality. The analysis results provided the quantitative effect of each of these factors. During a production run, if the part quality drifts, the analysis results can help in adjusting the process to get the part quality back to where it should be. For example, if the tornado diagram shows that the holding pressure and cooling time have the most significant effect on the length of the part, then a contour plot with holding pressure and cooling time as variables may help the processor adjust either one or both variables and continue making acceptable products. The guesswork regarding what parameter should be changed is eliminated. This will also keep the process sheet clean of changes to several parameters and limit the changes to the holding pressure and cooling time. Educating the workforce and providing access to this data at the molding machine is essential.

8.10.4 Setting Process Change Tolerances

The tornado charts and the contour plots provide information on the most significant process parameters and the extent to which they affect the part quality. Based on these data, the process change allowances and limits can be set. In the contour plot, a box, such as the one shown in Fig. 8.17, must be set and the corresponding limits must be used as tolerance limits for allowable changes during production. Referring to Fig 8.17, the tolerance for the holding pressure can be set between 41 and 54 bar. Only these factors must be changed in case there is a quality problem.

8.10.5 Setting Alarm Limits

Alarms are set on process outputs. Once the process tolerances are set as described above, parts must be molded within these limits and the process outputs should be recorded. For example, in Fig 8.17 for holding pressures of 41 and 54 bar, if the corresponding cushion values are 0.50 and 0.25 mm, the alarm limits must be set at these cushion values. If the cushion value exceeds 0.50 mm, it will mean that not enough plastic has entered the mold and therefore it is likely to result in a short shot or a part out of specification so that the machine will set off the alarm. If the cushion value is below 0.25 mm, this may mean that there is too much plastic in the mold producing a part with flash or a part out of specifications.

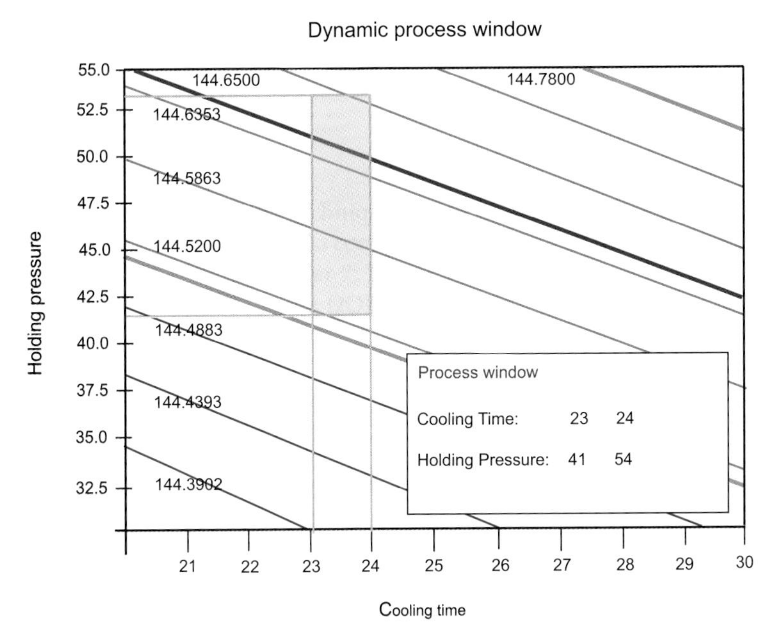

Figure 8.17 Generating the operating process window and setting tolerance limits

8.10.6 Reducing Inspection

The results from a DOE can be very useful in reducing and even eliminating inspection. If the process sensitivity charts indicate that a dimension does not seem to be affected by the various process changes and provided that this dimension is within specifications, it can be checked at mold startup and then be assumed as within specification during the in-process inspection. The same holds true for other dimensions that are well within the process windows of the established process.

Suggested Reading

1. Lahey, J.P., Launsby, R.G., *Experimental Design for Injection Molding* (1998), Launsby, Colorado Springs, USA
2. Del Vecchio, Understanding Design of Experiments (1997), Hanser Publications, Munich, Cincinnati

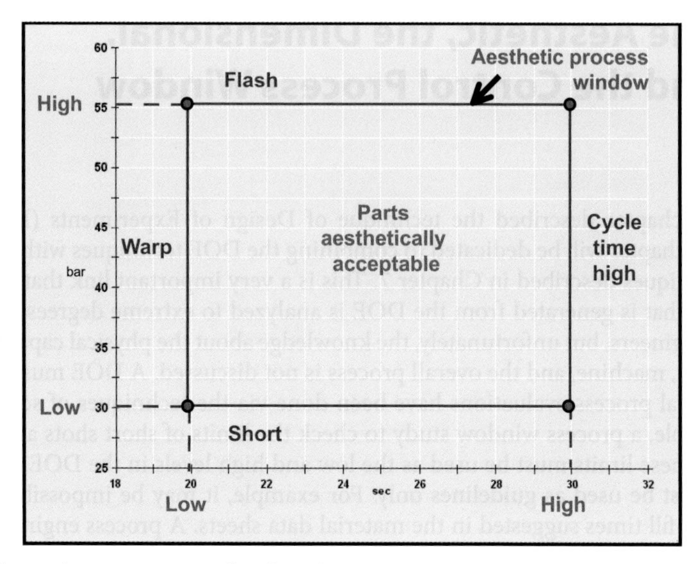

Figure 9.1 The aesthetic process window (APW)

X-axis: Cooling Time. Below 20 seconds, thicker areas of the parts appeared soft when ejected and there was evidence of warp. Therefore, this was taken as the low limit for the cooling time. Above 30 seconds, the cycle time became prohibitively long and therefore this was defined as the upper limit for the cooling time.

Y-axis: Holding Pressure. Below 30 bar, the parts had sinks and above 55 bar the parts flashed. Therefore these values were taken as the lower and upper processing limits for the part.

Within these limits the molded parts were acceptable aesthetically. They were free of all sink or flash. The DOE was performed using limits and a contour plot was plotted. This plot is shown in Fig. 9.2. The LSL, USL and the nominal values are also shown.

Table 9.1 Quality Requirements

	Length	Diameter
Nominal	144.65	6.18
+ Tolerance	0.13	0.09
– Tolerance	0.13	0.13
Upper Specification Limits (USL)	144.78	6.27
Lower Specification Limits (LSL)	144.52	6.05

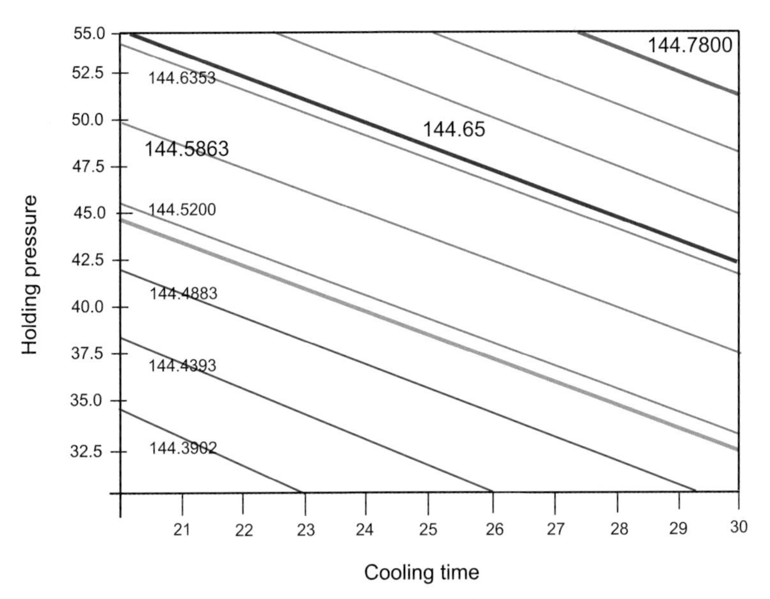

Figure 9.2 Contour plot for the 144.65 mm dimension in Table 9.1

9.2 The Dimensional Process Window (DPW)

The green shaded area in Figure 9.3 defines the process window in which dimensionally acceptable parts are molded. In this example, pressures above 800 psi will cause flash in the parts and below the red LSL line the parts will be out of specification. The window that is now a subset of the aesthetic process window is called the "Dimensional Process Window". In the case described, the window is not a uniform quadrilateral and the molding parameters must be selected inside this area.

9.3 The Control Process Window (CPW)

We shall now expand on the concept of dimensional process windows. If the dimension under consideration is a dimension that needs to be statistically capable, then based on the calculated control limits, a Control Process Window can be defined. The control limits are calculated on the standard deviation of the measured dimension. With the capable process, these limits are always within the USL and LSL and therefore the CPW is always within the DPW. The CPW is a subset of the the DPW. Once the molding process is started, the process capability can be calculated and the control limits can be established. Based on the statistical process capability, the Lower Control Limit (LCL) and the Upper Control Limit (UCL) were

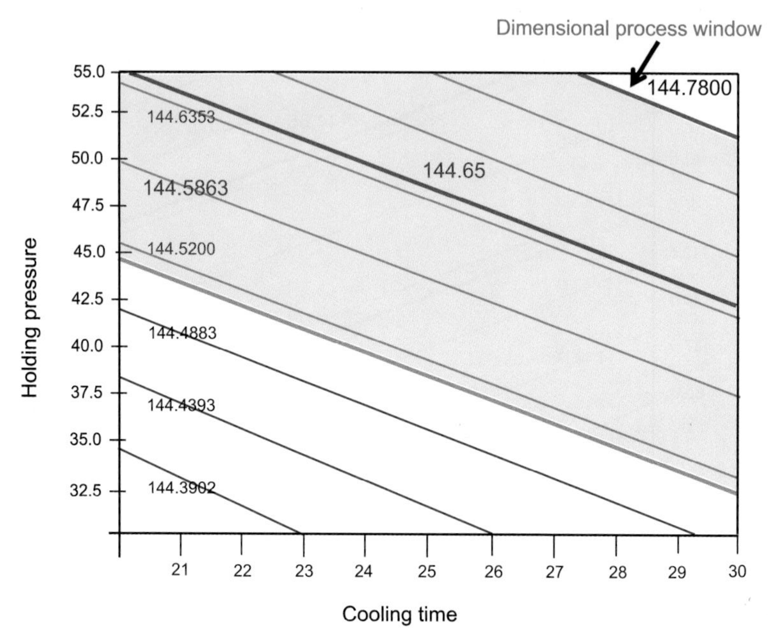

Figure 9.3 Dimensional process window for the 144.65 mm dimension in Table 9.1

calculated and are shown in Figure 9.4. These contours can now be plotted on the contour plot and the CPW can be determined. This new window, in which the parts are not only acceptable but also under statistical quality control, is called the Control Process Window (CPW). In Fig. 9.4 the yellow shaded area is the CPW.

With the explanations given in the preceding paragraphs a more precise definition of the terms are given below.

- Aesthetic Process Window (APW): The limits between which an aesthetically or cosmetically acceptable part can be molded. Dimensions are of no concern.
- Dimensional Process Window (DPW): The limits between which a dimensionally acceptable part can be molded.
- Control Process Window (CPW): When the statistical control limits are applied to the dimensions given by the dimensional process window, the resulting window of operation is called the control process window.

During the process of initial mold sampling the APW should be as wide as possible. The parts that are molded should be aesthetically acceptable over a wide range of processing parameters. Once a wide APW has been determined, we can use the low limits and the high limits to set the high and low levels required for each DOE factor. The higher the difference between the high and the low limit, the higher will be the magnitude of change on the quality of the part. The results of the DOE can then be analyzed and the DPW can be determined. The wider the DPW, the more robust the process will be. If the DPW is skewed towards the corners or the sides of the APW, an effort must be made to make steel changes in the mold

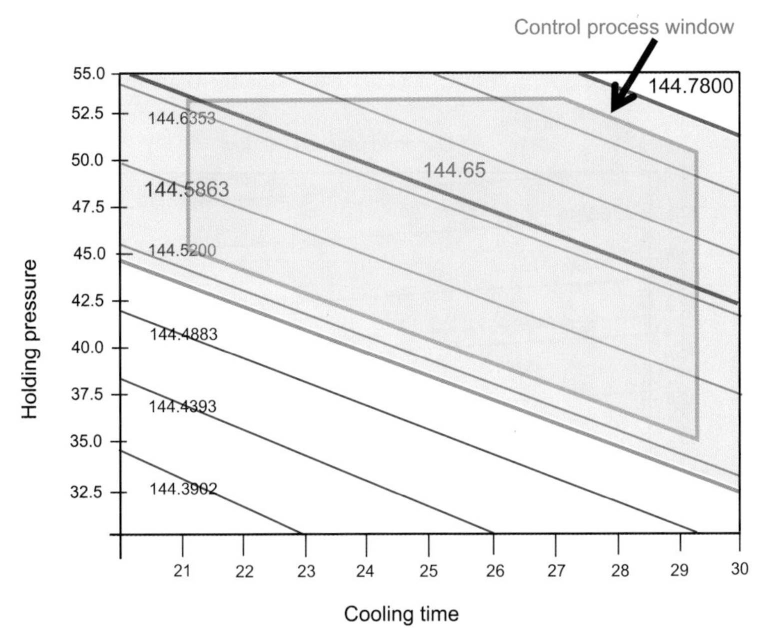

Figure 9.4 Control process window for the 144.65 mm dimension in Table 9.1

cavity to bring the nominal of the dimensions to the center of the APW in order to make the process robust. In an ideal situation, the DPW should be as big as the APW and both should be as wide as possible. The DPW will also give the product designer a chance to review the robustness of the dimension under consideration. Setting the tolerances to practical limits is often a battle between the product designer and the processor and the DPW is a good tool to facilitate such discussions.

The CPW is a subset of the DPW and the DPW is in turn a subset of the APW. Naturally, a wide APW favors the possibility of a wide DPW and CPW, provided the nominal of the dimension is centered in the APW. This data can be reviewed with the tooling engineer to support justifying a mold steel adjustment to center the part dimension within the specification limits which will provide a wider DPW.

9.4 Multiple Dimensions

The above discussion covered a single dimension under consideration; however, almost all parts have multiple dimensions. With multiple dimensions involved, the contour plots become exceedingly complex. The contours for the nominal and specification limits of each dimension will rarely overlap and it would be even more unusual for the slopes of the contours of each dimension to be identical to each other. The effective process window will therefore be the intersection of the two individual process windows. Figure 9.5 shows the control process window for another dimesion on the part whose nominal value is 6.18 mm. Figure 9.6 shows the overlap of the two individual APWs. The overlap is called the "Composite Dimensional Process Window "; it is typically smaller than the individual process windows.

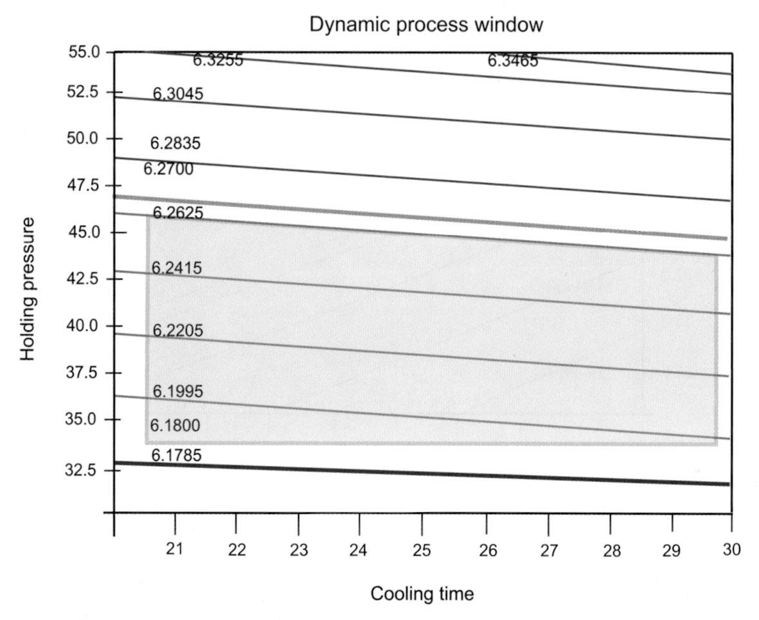

Figure 9.5 Control process window for the 6.18 mm dimension

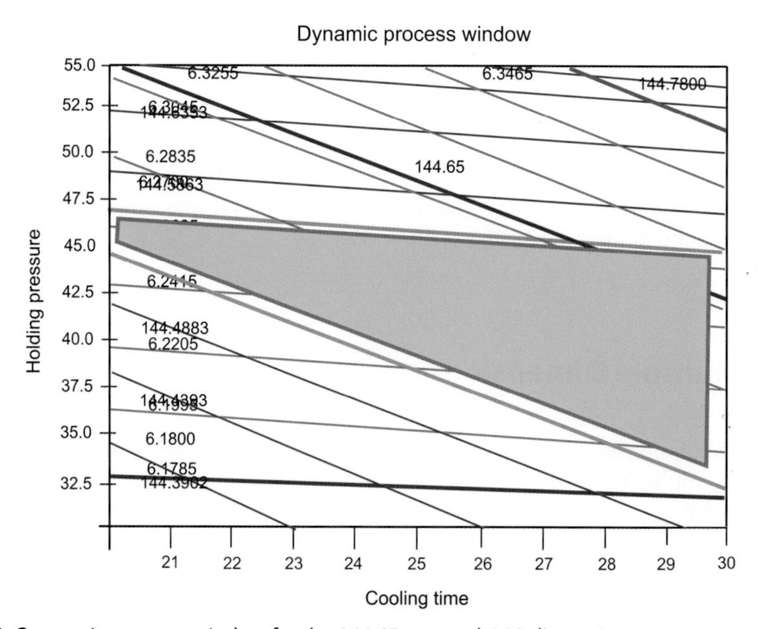

Figure 9.6 Composite process window for the 144.65 mm and 6.18 dimension

9.5 Multiple Cavities

The concept of multiple dimensions explained in the previous section can also be extended to multiple cavities. In theory, all cavities should be producing parts with the same quality. If that is the case, it is a matter of applying the results of one cavity to the rest of the cavities. Going back to the concept of cavity balance, if all the cavities fill evenly and the mold temperatures are even across the cavities, then the parts molded from these cavities will be near identical in part quality. Therefore, analyzing one cavity and verifying the results using all the cavities could be sufficient. Since all the cavities are subjected to the same melt temperatures, pressures and times, they should be identical. This is another reason why cavity balance is important. The knowledge of all cavities being identical can again lead to reduced inspection and analysis.

In case there is a difference between the different cavities, all dimensions of all cavities must be considered on an individual basis and analyzed separately.

9.6 Closing Remarks

Some would argue that the DPW and CPW gets smaller and smaller as the number of dimensions and cavities increase. Unfortunately that is the reality and in fact typically it is not acknowledged because of the effort involved in making the process windows large. A small process window leads to a process with almost no room for any adjustment, resulting in a process sheet with very small limits for process changes. Tolerances on process parameters are usually established based on past molding experience. However, to be successful, process tolerances must be set based on the type of analysis described in this chapter. This further pleads the case for well built injection molds and parts dimensioned with reasonably defined tolerances. This is where concurrent engineering principles and practices become important. A robust process requires less human intervention, which allows the potential for reducing the frequency of in-process inspection. A well planned injection mold and part design, along with a disciplined process development approach are critical to having an efficiently running production process and a profitable manufacturing operation. Without the above, much time, money, and resources will be wasted in trying to efficiently produce parts from an inherently inefficient system.

10 Mold Qualification Flowchart, Production Release, and Troubleshooting

This chapter will discuss a flowchart that includes all necessary measurements and the documentation that must be carried out for successful release of a molding process into production. The tools required for mold qualifications and troubleshooting are also mentioned. Development of troubleshooting guides is a continuous process and they need to be revised and updated during production runs.

10.1 Mold Qualification Flowchart

Process developmental procedures described in the previous chapters are key to releasing the mold to production. A robust process needs little or no supervision during the production run. The required supervision usually involves for the processor to perform a visual check on the product, verify the process, answer any alarms, or perform scheduled or preventive maintenance such as cleaning the mold faces. If processes are not robust, a technician needs to constantly attend the machine to adjust the process parameters in order to yield acceptable parts. Robust processes yield consistent parts, shot to shot, cavity to cavity, and run to run.

The mold qualification procedure can be split into two parts, the mold-function qualification and the mold-part qualification. The flowchart is shown in Fig. 10.1.

10.1.1 Mold Function Qualification Procedure

Mold function qualification requires a 6-step qualification study described in Chapter 7. During this study, the function of the mold and its components, the determination of some of the process parameters, and the size of the process windows are evaluated. During this study, dimensional analysis to measure process capability are typically performed. The actual dimensions are not of major importance because the next step is to adjust the mold steel dimensions to mold the parts within the required specifications. The mold function qualification step is to determine the aesthetic process window (APW). Naturally, a wide window is desirable. Once the mold function and process windows are acceptable, the next step will evaluate the part quality.

10.1.2 Mold and Part Quality Qualification Procedure

This step involves performing the DOE, selecting the process, and determining the size of the dimensional process window (DPW) and the control process window (CPW). At this point, based on the knowledge from initial statistical dimensional analysis, there must be

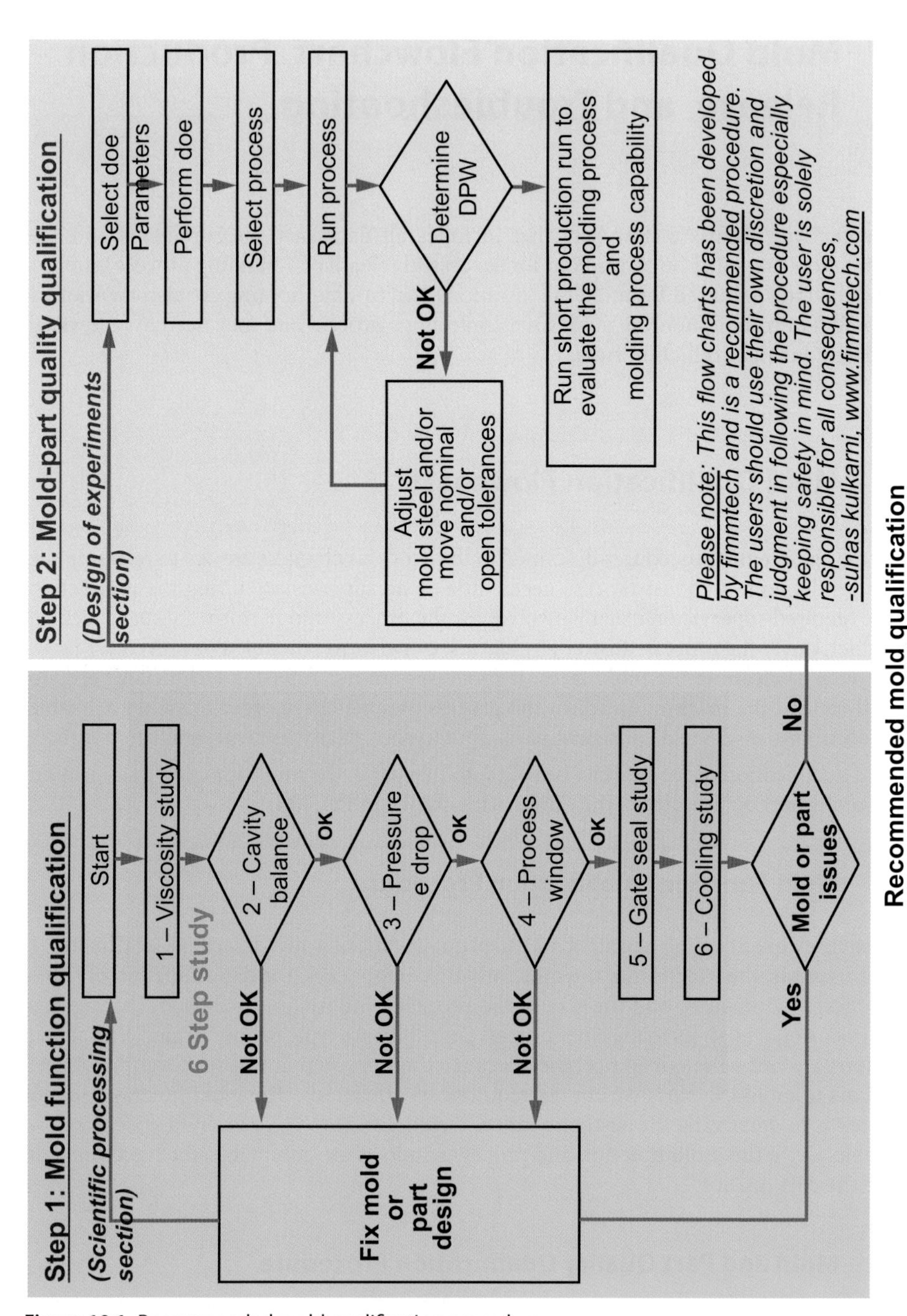

Figure 10.1 Recommended mold qualification procedure

enough confidence that the process is stable and that the mold steel can be and must be adjusted to center the process in the DPW and thus in the CPW. At this point, if the dimensional variation for a particular dimension is greater than the difference between the Upper Specification Limit (USL) and the Lower Specification Limit (LSL), it becomes clear that it is impossible to consistently mold all the parts within specifications because some parts will always be out of specifications. The product designer must revisit the product specifications and the material selection for the parts. It is acceptable to stop the mold qualification process at the end of step 1 (the mold function qualification), if the dimensions are acceptable. However, the only way to know whether the process has been optimized is to perform a DOE. The benefit of performing a DOE outweighs the time and effort required to perform it. Process tolerances, alarm limits, and process robustness are some of the critical pieces of information available only through a DOE.

10.2 Mold Qualification Checklist

Having a mold qualification checklist makes sure that all features of the mold and the process have been assessed. A sample checklist is included in Appendix F. The checklist should be used during the mold trial and should be completely filled out with comments by the end of the trial. Any suggestions must be passed on to the mold maker or the concerned department, including reasons for any recommended or required changes. For example, if the mold is flashing, a sample part with flash must be given to the mold maker. An example of a completed mold qualification checklist is shown in Table. 10.1.

Table 10.1 Mold Qualification Checklist

No.	Question	Comment
1	Are gate sizes acceptable ?	Open gate size to 0.060”
2	Are runner sizes acceptable?	Yes
3	Is venting acceptable?	Vents on runners required
4	Is part fill acceptable?	Yes
5	Is the ejection acceptable?	Parts do not fall off the ejector pins
6	Has the rheology study been done?	Yes
7	Is the cavity balance acceptable?	No, Cavity 2 is shorter than the others
8	Is the process window acceptable	Process window is small. Part flashes soon after fill
9	Has the gate seal study been done?	Yes. Will be repeated after gate size change
10	Has the cooling study been done?	No

10.3 Process Documentation

Because there are a number of factors that affect the process, a detailed record must be kept of each of these factors. The factors include machine settings, actual process outputs, material drying parameters, machine setup instructions, mold setup details, operator work instructions, secondary processes, and any other factors involved in the journey of the plastic pellet until it is shipped out of the facility as a finished product.

10.3.1 Process Sheet

The process sheet is usually the first piece of documentation that is generated during the initial sampling of the mold. It primarily contains the machine settings for the process. It should also include other factor-setting information. A typical list of factors is included in the sheet shown in Table 10.2.

Table 10.2 Process Variables and Outputs to Be Recorded on a Process Sheet

Temperatures	Speeds	Pressure	Times	Other	Outputs
Drying	Injection	Injection	Drying	Shot Size	Part weight
Barrel	Mold open	Packing	Packing	Transfer position	Runner weight
Hot runner	Mold close	Holding	Holding	Material info	Injection time
Mold	Ejection	Back	Cooling	Nozzle length	Cushion
Fixtures	Screw	Tonnage		Nozzle orifice	Cycle time
Annealing				Coolant flow rate	Melt temperature
				Core sequences	Screw recovery time

A process sheet must be generated and updated at the end of every mold trial iteration. Any new process changes that are made must be recorded over the old parameters, but a log of all the process changes with the corresponding dates must be maintained. This essential history is useful in debugging future issues. The history also provides information on the evolution of the process to the current state from "what it used to be". For example, over a few runs cycle times seem to drift away from the standard and the question always asked is "What changed?" A process change log is very useful in such cases and the changes over time can be evaluated. A process change log must also be maintained during production runs and should be a part of the process documentation.

10.3.2 Waterline Diagrams

Injection molding is a heat transfer process and the mold is the primary heat exchanger. The mold temperature plays a very important role in the part quality because it affects the rate of heat transfer. Mold temperatures must not only be kept constant during a run, but also must be maintained at the same values from run to run. This will help to achieve the run-to-run consistency goal. This leads to the necessity of hooking up cooling waterlines in the same manner for every run. The Ins, Outs, and the loops must be placed in the same place during every run to ensure the same heat transfer from the mold. A water line diagram must be maintained in the records, showing the hookup for every waterline. An example of a water line diagram is shown in Fig. 10.2.

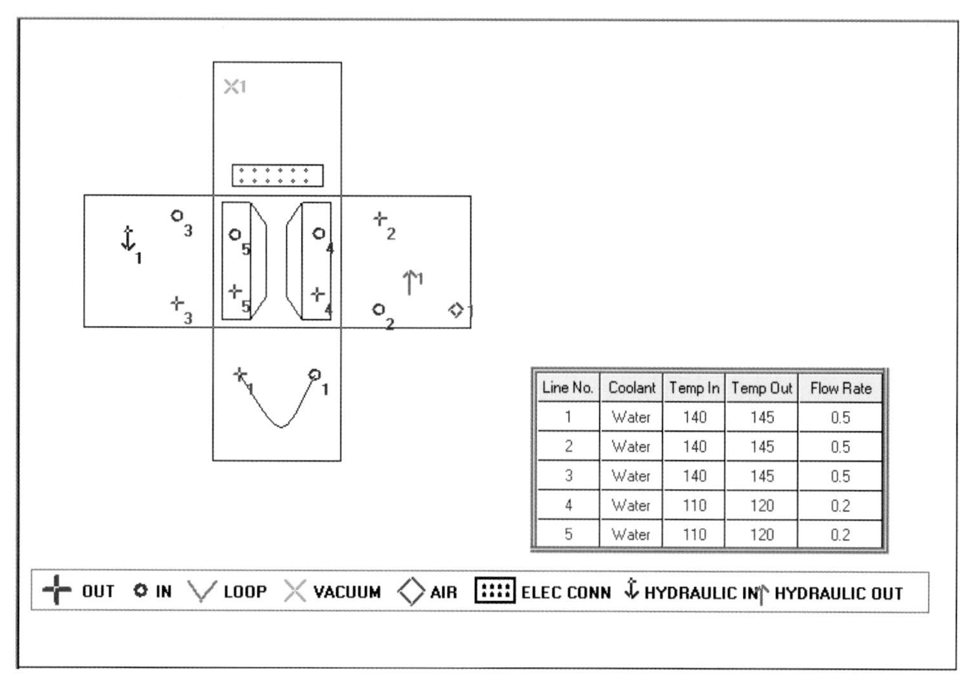

Line No.	Coolant	Temp In	Temp Out	Flow Rate
1	Water	140	145	0.5
2	Water	140	145	0.5
3	Water	140	145	0.5
4	Water	110	120	0.2
5	Water	110	120	0.2

Figure 10.2 Waterline diagram

Because the heat transfer also depends on the coolant temperature and flow rate, these values must also be recorded. The differential pressure or the pressure drop is useful in evaluating whether there is any scaling or other obstruction in the water lines. This information should typically be recorded by the tooling department.

10.3.3 Mold Temperature Maps

Even when the water lines are hooked up properly as documented in water line diagram, the actual mold temperatures are also important. For example, if the setting on the mold temperature control unit is 60 °C, but the water line is obstructed, the mold will not reach the required temperature. There is also a loss of heat between the temperature control unit and the mold. Water leaving the temperature control unit at 60 °C may cool down to a temperature of 50 °C by the time it reaches the mold. Therefore, the actual mold temperatures of the mold cavities must be recorded before startup. The temperature at each of the water fittings must also be recorded. Any plugged line will cause a different OUT line temperature than the rest of the lines. When the mold temperature is set close to ambient room temperature, a plugged line will not negatively effect the temperature distribution in the mold cavity, because the mold temperature will always find equilibrium in the cavity steel. It is therefore important to record and check the mold temperature when the mold has been running for some time. This time must be specified and recorded. Examples of mold temperature maps are shown in Fig. 10.3.

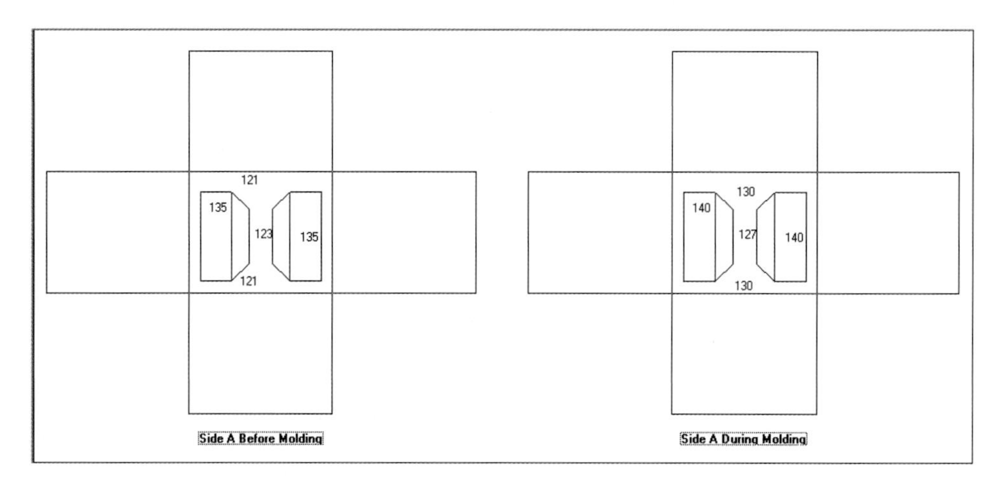

Figure 10.3 Mold temperature maps – before and during molding

10.3.4 Setup Instructions

The procedure and the tools required to set up the mold must be recorded to facilitate a quick and hassle-free set up. This is particularly important for molds that need special instructions, such as mold open and close sequences for a mold with cores. Molds can easily get damaged if the sequences are not correct, costing the company sometimes thousands of dollars for the components to be replaced, the manpower, machine time, loss of production, and so on. Mold start-up instructions must also be included. A mold must be started up by lowering both pressure and speeds. Mold shut-down procedures and material purging instructions must also be provided. Times for heat soaking of hot runner manifolds must also be provided for molds that utilize a hot runner system.

10.3.5 Operator Instructions

Packaging and any secondary operations performed by an operator must be clearly defined. The operator is usually the last person to handle a part before it goes into the box, making him/her the integral part of all quality control efforts. Operators should be given clear instructions regarding handling the product and performing any secondary operations, such as deflashing the product. Packaging is another important step, because parts not packaged correctly could become damaged during transportation and in some cases this could result in a change in quality, such as warpage.

10.4 Documentation Books

It is best to have two types of documentation records, The *mold qualification book* and the *production book*. The mold qualification book should record all details of the qualification process and the production book must include all information required for production, including some of the qualification results to help the technician debug and fix any problems during production. Table 10.3 provides a list of documents that should be included in these books.

10.5 Qualification Production Runs

Once part quality, process windows, and process capability are acceptable, a short qualification run must be performed. The length of this run should be determined based mainly on the production volumes and the criticality of part quality. For molds that would typically run for most of the year, a robust process and mold design will relieve production personnel of constant production management and will allow running a lean manufacturing operation. For such molds a qualification run of 24 hours should be considered. For critical parts, such as parts that require zero defect quality, the same procedure must be followed. It is extremely important that the process is closely monitored during this production run and process changes must be avoided. If process changes are required to maintain the quality, the whole process must be reevaluated, because this is an indication of a non-robust process. The parts molded from these runs must be initially quarantined and released to the customer once all the information and quality has been verified. Longer runs will provide a better understanding of the dimensional variations.

Table 10.3 Documents to Be Included in Qualification and Production Books

Document	Qualification book	Production book
Part drawing	Y	Y
Properties data sheet	Y	N
Processing data sheet	Y	Y
Mold setup instructions	Y	Y
Viscosity curve	Y	Y
Cavity balance	Y	Y
Pressure drop	Y	Y
Aesthetic process window	Y	N
Gate seal study	Y	Y
Cooling study	Y	N
DOE matrix	Y	N
DOE results – Pareto charts	Y	Y
DOE results – contour plots	Y	Y
DOE results – other	Y	N
Process sheet	Y	Y
Dimensional process window	Y	Y
Control process window	Y	Y
Process change log (during development)	Y	N
Process change log (during production)	N	Y
Operator instructions *	Y	Y
Daily mold maintenance (during production)*	Y	Y
Start-up and shut-down instructions*	Y	Y

* These are living documents that should be started during mold development and must be constantly updated during production

10.6 Mold Specific Troubleshooting Guide

Even for a commonly known defect, the solution to fix the problem may be different for each mold. For example, splay on one part could be a result of excessive injection speed, while for another mold running the same material, the splay could be the result of low mold temperature. Every mold has its own characteristics and therefore, although a general troubleshooting guideline is a good starting reference point, every production book must contain a living document that contains the record of the typical problems and the respective solutions. This record should be updated every time a new defect is seen and a solution for the defect has been found, saving effort the next time the defect is seen. The factors that must not be changed must also be mentioned. For example, if mold temperatures are important for a particular dimension, or if it was found not to be a significant contributor to the solution of the problem, then this must also be mentioned. Often, there is more than one solution to fix a defect and therefore the solutions must be prioritized. The results from the DOE must also be included here. They will give an indication of the most important factors that have an effect on the parts. An example of a trouble shooting guide is shown in Table 10.4.

Table 10.4 Mold Specific Troubleshooting Guide

No	Defect	Recommendation	DO NOT CHANGE
		Clean vents	Injection speeds
1	Splay	Check material moisture	
		Reduce melt temperatures (not below 210 °C)	
2	Shorts on the sleeve	Clean vent pin	Holding pressure
3	15.055 dimension is undersized	Check water flow through the cores	Holding pressure

10.7 Molding Startup and Shutdown

10.7.1 Purging

The barrel must be purged off all the material present in it to eliminate any purging compound that was used and/or eliminate any degraded material present in the barrel. Injecting degraded materials in the mold can cause two potential problems. First, the degradation could have been so significant that all mechanical properties of the plastic have been lost and injecting this material in the mold can produce a brittle part, posing ejection problems. Parts getting stuck in the mold, parts breaking off during ejection, ejector pins piercing through the parts, breaking of ejection components due to stuck parts are all problems caused by degraded materials. The second problem that may arise is the excessive gassing that takes place due to the material degradation. The gasses build up in the barrel and if not purged, they get injected into the mold and again end up clogging the vents that are the last points to fill. The products of degradation are easy to transport through the mold and clog the vents. If the vents get clogged during the first shot and this problem is not noticed and fixed, the molded parts may develop cosmetic and dimensional problems immediately. This forces a process change further effecting parts and finally leading to a chain of process changes.

Hot runner systems must also be purged of possible degraded materials using fresh material. A large piece of cardboard must be placed on the moving side of the mold to avoid any degraded material from being sprayed on the mold, because if this happens, the mold often needs to be pulled out for cleaning, especially if there are complicated slide or lifter mechanisms. Care must also be taken to inject at lower speeds and pressures, because some hot runner manifold components rely on the clamping pressures to keep them from leaking under injection pressure. A good starting method is to use the screw speed and a high back pressure to extrude the melt out of the hot manifold. All molds are different and can pose unique problems with regard to purging. Each purging procedure must be evaluated for the particular mold before being carried out.

10.7.2 Startup of a Molding Machine

All molding operations involve high pressures, temperatures, and speeds. These can be extremely dangerous and accidents, from burns to fatalities, have been recorded. Safety should always be a priority and starting up the molding machine must be done following a careful and systematic procedure. From the process standpoint, it is preferable start up the molding process using a conservative approach, detailed in the following, before full pressures and speeds are set on the molding machine. This will prevent any accidental damage and over-packing of the parts in the mold. Over-packing of the parts causes flash and in some molds the flash can fill up internal vents. Stationary vent pins are examples of such internal vents. In addition, when the parts get over-packed, mold opening pressures increase and it becomes difficult to open the mold.

To better illustrate the startup procedure, here is an example with the following conditions: Holding pressure = 1000 psi, holding time = 8 seconds, and cooling time = 20 seconds.

Based on these conditions, a recommended startup procedure would be:

1. Verify the process with the documentation that was recorded during the qualification run
2. Check melt temperature
3. Check mold temperature and make sure water is flowing
4. Increase cooling time by an amount equal to the pack and hold time (cooling time = 28 s)
5. Set pack and hold, pressure, and time to zero (HP = 0, HT = 0)
6. Set the screw delay time equal to the set HT: Screw delay = 8 s

7. Purge the machine and take the first shot in a semi-automatic mode
8. Match the 'Injection Only' part with the recorded documentation or sample part
9. Take a couple more shots
10. Add half the hold and pack time and pressure and reduce the cooling time by half the set HT (HP = 500 psi, HT = 4 s, cooling time = 24 s)
11. Take about 5 shots
12. Set screw delay time equal to zero or the value in the recorded documentation
13. Set the process to match the recorded documentation

Note: Some parts cannot be molded short. In such molds, leave the holding time to the set value and gradually increase the holding pressure.

10.7.3 Shutdown of a Molding Machine

The shut down procedure is equally important. If a heat sensitive material or an easily degradable material is molded, it must purged from the barrel before shutting the machine down. Molding must be stopped and an appropriate purging compound must be used to purge the barrel. If the mold has a hot runner system, follow the recommendations of the hot runner supplier to purge the hot runner system before shut-down. All material in the barrel must be emptied out and the screw must be left in the front position. If the material is not emptied out and the screw is not left in the forward position before being shut down, a solid cylinder of plastic will be formed once the machine has cooled, which will pose a problem during start-up.

Because plastic is a bad heat conductor, the center of the cylinder takes a long time to melt; therefore, the solid cylinder prevents any purging or injection from taking place. When chilled water is run through the molds, the water is shut off and a few shots are run to prevent water condensation in the cavities.

The last shots with the runners must always be saved with the mold. No secondary operations such as degating or deflashing should be done on these shots because they are evidence of the condition of the mold and process before shut down.

10.8 Troubleshooting

Troubleshooting is probably the most important job function in any production environment. As equipment and processes have become more and more complex, the need for highly knowledgeable workers with good troubleshooting skills has become increasingly important. The molding process is complex in terms of plastic flow characteristics and the speeds, pressures, times, and temperatures involved with the process. Each of these factors can affect the quality of the part. For example, increasing melt temperature can increase flow rate, or increasing injection speed can also increase flow rate. Increasing cooling time can increase the dimension of the part, while decreasing the mold temperature can also increase the dimension of the part. Because multiple actions can have similar effects on part quality, different people have their own preferred method of solving a problem. The downside of this fact is that the process sheet is often updated with the latest changes and after a few months of running the mold, the set process ends up being completely different from when it was first created. This illuminates two problems: first, the process was not established scientifically and second, the personnel were not trained to troubleshoot the process in a systematic manner with the right tools. But even with a robust process, often production problems may still occur. When a robust process is established, documenting the complete process with all inputs and outputs must be done; however, it is sometimes impossible to capture the status of each related process or equipment. For example, recording the response time of a hydraulic valve controlling the injection or a pneumatic air valve controlling the gate is difficult and probably never done, although this may have an impact on the part quality. If the conditions of these components change, so will the molded part quality requiring troubleshooting skills.

There are no hard and fast rules for troubleshooting an injection molding process. First, the problem with the parts must be understood and then the possible reasons for the issue must be considered. No process parameter must be changed until the complete set of process parameters is compared with the original recorded process parameters and the outputs are scrutinized. Every other factor that is not recorded on the process sheet must also be looked into.

The following are some guidelines that can be followed for general troubleshooting.

1. Compare the set process to the original documented process. If there are any changes, do not change the process back to the original (because they may have been made for a reason).
2. Observe the whole process for ten shots and record all the outputs for these ten shots. These include: injection fill time, cushion, screw recovery time, and cycle time. Compare the above actual outputs to the documented ones.
3. Stop the machine and compare the actual mold temperatures to those that were documented in the book.
4. Record the melt temperature and compare with the documented measurements.
5. Check the temperature of the hoses to check for water flow as described earlier. Remove any obstructions in the lines and let the water circulate for some time to allow the mold temperature to stabilize.
6. If there have been any changes compared to the original process sheet, refer to the results from the DOE and see if those changed parameters would have any effect on the quality of the part. If not, set the process back to the original. If the changes are influential to part quality, check if the out-of-specification part was a result of this change made. If so, change the process parameter back to the original. If not, make the appropriate change and bring the parts back into specifications. Any process changes must be recorded.

Once the parts are running meeting quality requirements, the process and quality must be constantly monitored for about a shift's worth of production. Any unusual trends, such as increasing injection pressures, decreasing fill times, or any other variations must be noted. This change may sooner or later cause the parts to go out of specifications and one must be proactive to find the root cause of the trend or variation. A good example is inadequate cooling of the mold because of lack of cooling lines. In its role as a heat exchanger the mold must remove all heat that is put into it to maintain steady state. If the heat is not removed, the mold temperature will rise and affect the quality of the part. This change will be gradual and will be manifested in a certain trend in the dimensions of the part.

A process change log sheet must be maintained with the production book. The original process must always be present in the book, together with the updated process sheet. Every time a change is required, the old and new process parameters must be recorded on the process change log sheet. This disciplined record keeping will ensure that a history of all changes made to the process sheet is maintained for review.

Checking water flow through the mold is not a very easy task. When the mold temperature control water unit is turned on, it is assumed that the water is flowing through each line in the mold. One of the most common reasons for dimensional issues in production is that a water line (or oil line) becomes obstructed. These issues can be detected only by physically checking the temperature of the hose with a pyrometer, if the water temperature is high. For temperatures between approx. 30 to 45 °C (85 to 115 °F), the most efficient way is to hold a hose by hand and feel the heat of the water flowing through the lines. For water temperatures below 30 °C, this technique can be confusing because the temperature is not very different from room temperature and holding the hose by hand and trying to feel for a warm hose may not indicate water flow. Machine vibration can easily be confused for water flow and therefore trying to 'feel' the water flowing through the mold must be avoided. It is good practice to raise the temperature to approx. 35 °C and check for the water flow. Safety must be practiced. The practice of checking for water flow must also be followed at the start of every job. The water must be set to approx. 35 °C, the water flow must be checked, and then set to the required process temperature. Certainly for molds that use oil as a coolant a pyrometer must be used because the mold temperatures can be very high depending on the material. For example, PEEK is processed at mold temperatures between 175 °C and 205 °C (350 °F to 400 °F). The importance of safety cannot be stressed enough here because of the high temperatures of the coolant oil that is used in processing such materials.

Below are some of the common problems encountered on the production floor:

1. Pinched or kinked waterlines
2. Water line control valve not turned on
3. IN and OUT reversed
4. Inlet volume of water to the manifold is less than the required output to the machine
5. When using a T-junction for water lines, the IN must be such that it is the vertical line of the letter and the OUTs are the horizontal lines. If the water IN is one of the horizontal lines, the OUT on the vertical line can get starved of water. A venturimeter works on the principle that vacuum is generated in a vertical line of a T-junction if the internal geometries and sizes of the T-junction and the fluid velocities satisfy Bernoulli's equation. Although the generation of vacuum is unlikely, the effect of reduction of water flow is highly possible.
6. Water lines from the controller to the mold are too long, causing loss in temperature and increasing pressure drop
7. Feed throat temperature is not regulated

10.9 Important Equipment and Tools for Qualifications and Troubleshooting

Although this seems like a topic that does not need discussion, it is included here more as a checklist and to emphasize the importance of this equipment.

1. Melt pyrometer
2. Surface pyrometer
3. Weigh scale with the following accuracy (preferred)
   - 1 g for parts weighing over 250 – 300 g
   - 0.1 g for parts weighing between 50 – 250 g
   - 0.01 g for parts weighing less than 50 g
   - 0.001 g for micro-molded parts
4. Flow meter
5. Magnifying glass
6. Flash light and mirror
7. Brass tools – rods of various sizes (diameter and length), pliers.
8. Flame torch
9. Heat gloves – thin and thick
10. Processing data sheets
11. Camera – still and video
12. Calculator
13. Notepad, forms (with copies) and procedures of mold qualification

The use and benefits of most of the equipment is self-explanatory. The actual melt and mold temperatures must be measured with the pyrometers, rather than recording the settings. Accurate weigh scales are required to perform gate seal tests or to perform statistical analysis on part weights. Part weight is a valuable output that directly correlates to the quality of the part. If no variation is seen in the part weights, the weigh scale resolution is not small enough. An accurate scale with sufficient resolution to detect small changes in part weight is a necessity. A flow meter need not be installed on every single water line, but it should be used to check the water flow during the preventive maintenance schedules. Those numbers should be used to compare the flow rates during production, if there is a problem with part quality. When dealing with plastic or parts stuck in the mold, brass tools or those made of soft materials must be used to prevent damage to the mold. For highly polished surfaces, it is best that only experienced technicians handle any stuck plastic or any issues with the mold. The polished surfaces can easily get damaged and repairs are not only expensive but also time consuming. Video cameras can help with infrequent problem occurrence. In such cases, setting up and then reviewing the recording is an easy solution. The recording can be paused or played back in slow motion to pinpoint the issue. Since documentation is a large part of the whole qualification process, documents, such as data sheets and forms, should be readily accessible to make the process efficient. Laptops and workstations are becoming commonplace on the production floor and the need for paper documents is slowly decreasing.

Table 10.5 Troubleshooting Guide for Molding Defects

Defect	Possible reason				
	Process	**Material**	**Mold**	**Machine**	**Part design**
Short shot	• Low plastic pressure	• Low melt flow rate • Low melt temperature • Contamination	• Low mold temperature	• Max available pressure low • Restriction in nozzle	• Long flow length
Sink	• Low plastic pressure	• Low melt flow rate • Low melt temperature • Contamination	• High mold temperature	• Restriction in nozzle	• Long flow length • High part thickness
Flash	• Plastic Pressure	• High melt flow rate • High melt temperature • Contamination	• High mold temperature	• Low tonnage	• Thick sections with thin sections in vent areas
Splay	• High Injection Speeds	• High moisture levels	• Low mold temperature • Restriction in the flow • Small gate size	• Restriction in nozzle	• Flow thickness too small
Warpage	• Melt temps out of range (low or high) • Mold temp out of range (low or high) • Cooling time low • Packing pressures high or low	• Material selection incorrect	• Gate location incorrect • Cooling lines not adequate • Cooling lines improperly placed • Water flow through lines is laminar	• NA	• Non uniform wall thickness • Long flow lengths • Support features such as ribs absent

Table 10.5 (*continued*) Troubleshooting Guide for Molding Defects

Defect	Possible reason				
	Process	**Material**	**Mold**	**Machine**	**Part design**
Voids / Bubbles	• Low pack pressure • Inject speed high • Low back pressure	• Material not dry • Contamination	• Poor venting • Flow pattern causing trapped air pockets	• Back pressure inconsistent	• Part thickness high • Flow pattern causing trapped air pockets
Contamination/ black specs or streaks	• Melt temp high • Screw speeds high • Inject speed high	• Contamination with a another plastic of lower processing temperature • Contaminated regrind	• Sprue/runner and/ or gate size undersized	• Machine not purged after previous run • Worn screw and/ or barrel	• NA
Brittle parts	• Melt temps out of range (low or high) • Mold temp low	• Material contamination • Material overdried • Regrind pass number high	• Poor venting • Sprue/Runner and/ or gate size undersized	• Worn screw and/ or barrel	• Presence of sharp corners
Burn marks	• Melt temp high • Inject Speed high • Screw Speeds high	• Material contamination • Regrind pass number high	• Poor venting • Sprue/runner and/ or gate size undersized	• Worn screw and/ or barrel	• Presence of sharp corners
Dimensions out of specifications	• Process not set per recorded sheet • Process not scientifically established	• Material not dried • New lot properties out of specifications	• Restriction in sprue/ runner	• Check ring leaking • Machine not delivering correct pressures and speeds	• NA
Shot-to-Shot inconsistency	• Process not scientifically established	• Material contamination • Regrind pass number high	• Restriction in sprue/ runner	• Check ring leaking • Machine not delivering correct pressures and speeds	• Tolerances are narrow

10.10 Common Defects, Their Cause, and Prevention

Some of the commonly encountered defects are mentioned in the following, together with the possible solutions to fix these problems. Independent of the encountered problem, if the mold was producing acceptable parts at one point, the documented process must be revisited and compared to the status quo. The solutions provided in Table 10.5 are guidelines and recommendations to problem solving. The most common defects are mentioned first. Performing a DOE as a troubleshooting tool is always useful when the problem is not easily solved or the solution does not produce a consistent product.

11 Miscellaneous Topics Affecting the Process – Mold Cooling, Venting, and Regrind Management

11.1 Mold Cooling

The mold is basically a heat transfer unit. The hot plastic melt is injected into the cold mold cavity where it is cooled down until it is ejected out of the mold. The mold is kept at a lower temperature compared to the melt, which is sufficient to bring the melt down to the ejection temperature of the plastic. The temperature of the mold is very important because it controls the rate of heat transfer. The aim of scientific processing is to achieve three different consistencies: shot-to-shot consistency, run-to-run consistency, and cavity-to cavity consistency. Consistent heat transfer is important to achieve this goal. The rate of heat transfer between the melt and the mold is directly proportional to the difference in the temperature between the two. If melt and mold temperatures are held constant throughout the molding run, it is safe to assume that the rate of heat transfer was consistent during every shot. Melt temperatures are achieved by setting the barrel temperatures on the molding machine and typically they are not a major source of variation. It is always good practice to set alarms on the actual barrel temperatures to detect any changes and out-of-limit conditions. Although it is not common practice to have a melt temperature sensor inside a mold, this is an excellent way to measure the consistency of the melt temperature.

Maintaining a constant mold temperature is much more challenging compared to maintaining constant melt temperatures. This is where the mold design is critical. Cooling circuits must be designed for efficient heat transfer. The choice of the coolant must also be considered. The word 'cooling' is used because it is necessary for the mold to be at a lower temperature than the melt in thermoplastic injection molding. Mold temperatures can be as high as 162 °C (325 °F) for materials such as polyimides, with melt processing temperatures as high as 400 °C (750 °F). Therefore, a few important considerations regarding the design of cooling channels for molds are discussed in the following.

11.1.1 Number of Cooling Channels

Uniform heat transfer between every section of the melt and the mold steel will reduce the melt temperature evenly throughout the part. This will lead to consistent shrinkage, eliminating any warpage or built-in stresses in the molded part. The most ideal scenario is to have a constant steel temperature at every point in the cavity, which means that the coolant needs to envelope the cavity steel. Naturally, this is not possible and therefore an attempt must be made to place the maximum amount of cooling lines around the part. This is not an easy task

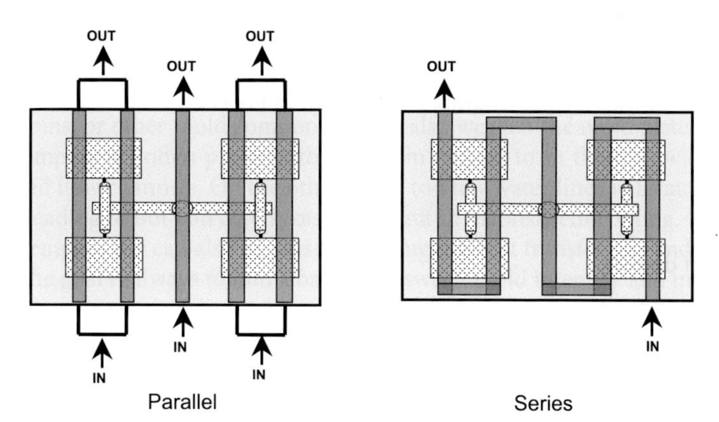

Figure 11.2 Parallel and series configurations for coolant lines

Parallel cooling lines exhibit less pressure drop allowing the coolant to flow at a higher velocity and therefore they pose less risk for laminar flow of the coolant through them. However, the flow through the lines may not be uniform. Because the coolant will take the path of least resistance, it is possible that a cooling channel with any restriction or blockage may not get sufficient coolant flow. It will be difficult to observe this without having a flow meter on every individual line. In case of a series arrangement, any restriction or blockage in a water line can be easily located; here, one flow meter is sufficient. If all molded parts suddenly start having quality problems, this could indicate a problem in a water line. Water lines hooked in series can exhibit large pressure drops, if the overall length of the line is too long.

11.2 Venting

The air inside the cavity needs to be evacuated to ensure plastic filling the cavity. Therefore, vents are added to the cavity blocks in the mold. Air that is not evacuated gets pressurized and super-heats, resulting in a dieseling effect on the plastic parts. This will cause the plastic to burn and/or create an unfilled area, causing a short shot. Over time, the mold steel can get damaged because of the excessive air pressures in a local area at the end of fill or in corners where the air and plastic tend to get pressurized. Figure 11.3 shows a part with burn marks before venting and without burn marks after venting. Figure 11.4 shows a rib that was short for lack of venting. Internal voids are another common defect that is seen with insufficient venting. Depending on the flow pattern of the plastic, the air gets trapped inside the part, forming voids. An example of a part with internal voids is shown in Fig. 11.5. Once the vents were added to the mold, the voids disappeared. Lack of vents can also create excessive pressures in the cavity, causing the mold to open sufficiently enough to cause flash on the parting line. Some machines have an option for mold 'breathing' before the start of the holding phase. The mold is allowed to open slightly to let the air out and then is clamped before the start of the pack and hold phase.

Figure 11.3 Elimination of burn marks after the addition of vents

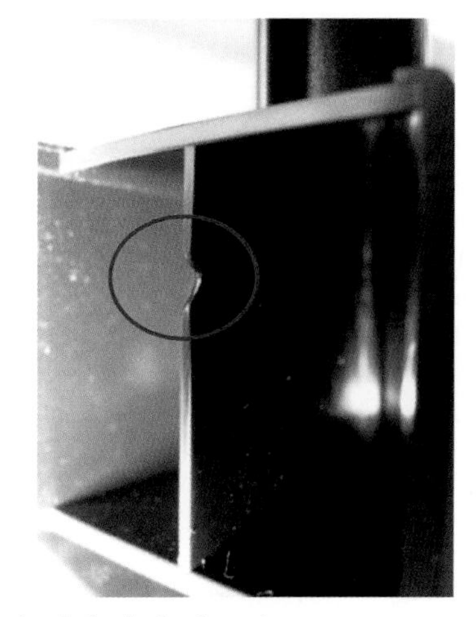

Figure 11.4 Rib not filled completely for lack of venting

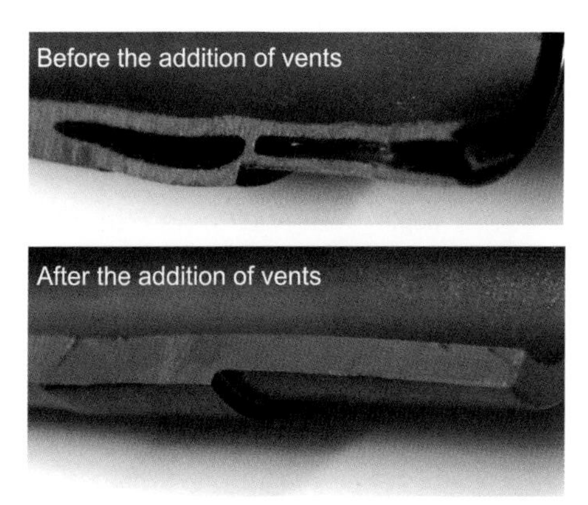

Figure 11.5 Elimination of voids after the addition of vents

11.2.1 Dimensions of the Vent

The vents connect the inside of the mold to the outside the mold. The viscosity of the plastic must be high enough to prevent it from flowing out of the mold through the vent. Figure 11.6 shows the cross section of the mold with the vent area. The relieved section that is closest to the cavity steel is the primary vent. The dimensions of the primary vent are the most critical. First, the vent depth should be such that it provides evacuation of air but prevents leaking of plastic material through the vent. Vent depths are discussed in detail in the next section. Second, the length of the vent land should not cause a pressure drop, preventing air from being pushed out, nor should it be so short that the plastic finds its way out to the secondary vent. Typical land lengths should range from 1.2 to 1.5 mm (0.060 to 0.080 in), assuming the vent depth is designed correctly. The vent width should be at least between 5 and 8 mm (0.200 in to 0.320 in) in size. On the high end, it can be as wide as desired and in some cases it can run around the entire perimeter of the part (ring vents).

The secondary vents are also called vent reliefs. The dimensions of the secondary vents are larger than the primary vent and therefore provide easy transport of the air inside the mold out to the atmosphere. Secondary vents should be approx. 0.25 mm (0.010 in). In all cases, the vents must be well polished to avoid any build-up of residue caused by the gases. The vents must also be draw polished in the direction of air flow. If the path of the air to the atmosphere is long, another step in the vents must be considered. The additional step could be as deep as 0.6 mm (0.025 in). These tertiary vents help in reducing the pressure drop and should also be draw polished in the direction of air flow.

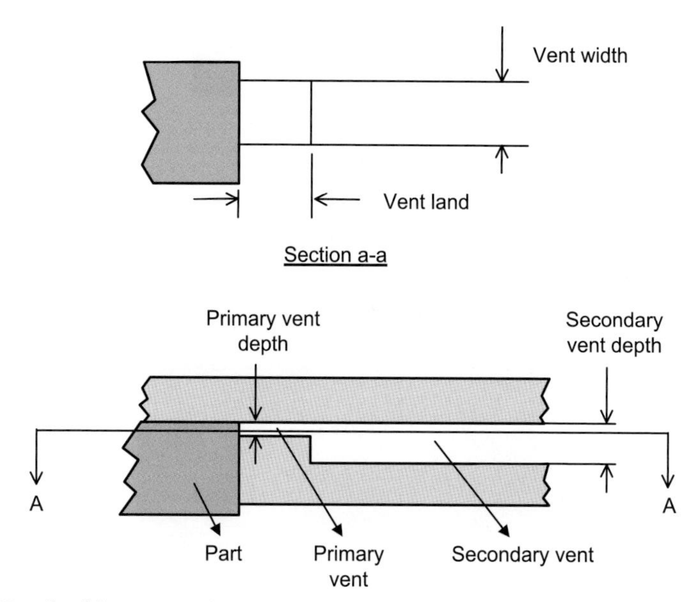

Figure 11.6 Details of the vent section

11.2.2 Primary Vent Depths

Vent depths depend on the viscosity of the plastic at processing temperatures and therefore differ from plastic to plastic. Typically, the material manufacturer will provide recommendations regarding vent depths. For example, the recommended vent size for ABS is 0.05 mm (0.002 in). Mold makers typically follow these recommendations when building an injection mold and tend to stay at the lower end of the recommendation to avoid flashing of the mold. If the plastic is able to enter the vent, this indicates that the vent is too deep and that it must be corrected. This involves welding and repairing the surface. To avoid this, mold makers are conservative and leave the vent depth steel safe. It has always been thought that it is the viscosity of the plastic that is the only factor important when determining the size of the vent. In a recent study it was found that the size of the vent depends not only on the viscosity of the plastic but also on the thickness of the plastic section behind the vent. To conduct this study, a special mold was constructed (see Fig. 11.7). The test part produced by this mold was center-gated and had 18 tabs of varying combinations of tab thicknesses and vent sizes, shown in Fig. 11.8. The center gate delivered the melt to each of the tabs at the same time. There were three tab thicknesses: 0.125 in (3.175 mm), 0.0625 in (1.587 mm), and 0.0312 in

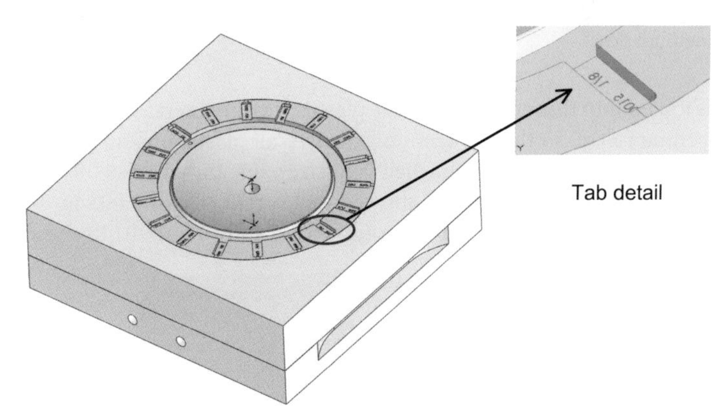

Figure 11.7 Vented mold for studying vent sizes

Figure 11.8 Test part molded with the vent test mold

(0.792 mm) and six vent sizes: from 0.0005 in (0.0127 mm) to 0.0030 in (0.0762 mm) in steps of 0.0005 in (0.0127 mm). Different materials were molded and for each tab the minimum vent size that produced flash was recorded. The test results indicated that it was the thickness of the tab that also played an important role in determining the vent size. The thicker tab was able to accept larger vents without flashing. For example, with an ABS material that was tested with a 0.125 in tab, the vents flashed at 0.0030 in vent size. On the other hand, the vents for the 0.0312 in tab, flashed at 0.0020 in vent size, see also Fig. 11.9.

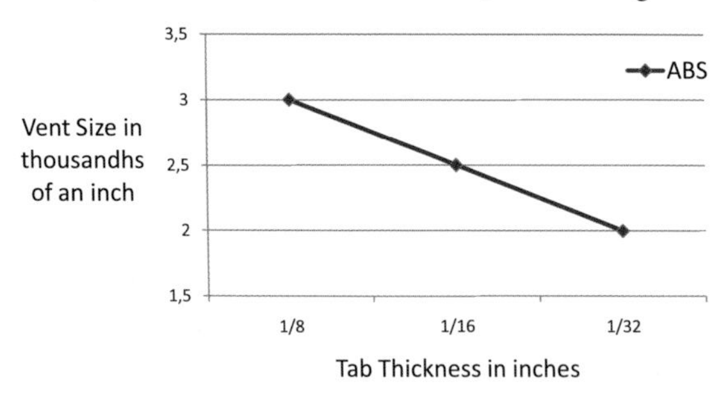

Figure 11.9 Graph showing the combination of tab thickness and vent depth at which flash begins to occur for ABS (medium flow)

As the plastic is forced into thinner sections, the plastic pressure increases, thus forcing the plastic out of the primary vent. In thicker sections, the overall plastic pressure is lower and therefore the vent sizes can be larger. The published vent size value for ABS is 0.002 in (0.0508 mm), but the test results from the vented mold showed that a vent size of 0.003 in (0.0762 mm) is acceptable. This value was used on several molds successfully. Figure 11.10 shows the various

Tab Thickness = 0.125"
Vent Depth = 0.0025"
No flash

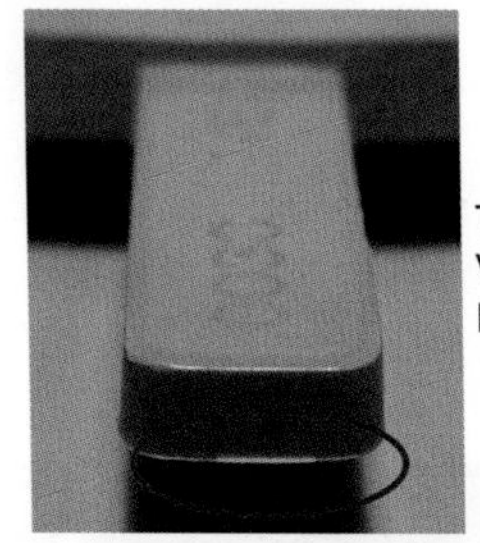

Tab Thickness = 0.125"
Vent Depth = 0.0030"
Evidence of flash

Tab Thickness = 0.03125"
Vent Depth = 0.0015"
No flash

Tab Thickness = 0.03125"
Vent Depth = 0.0020"
Evidence of flash

Figure 11.10 Tab/vent combinations showing no flash and evidence of flash

tab/vent size combinations that defined the start of noticeable flash in the vent. Tests on nylons also showed surprising results. The published vent size value for nylons is 0.0005 to 0.0007 in (0.0127 to 0.017 mm), but for thicker sections, a vent size of close to 0.0015 in (0.038 mm) could be used. In all cases it is important that proper processing techniques were followed. This again makes a case for following scientific molding procedures to establish robust processes.

11.2.3 Location of Vents

The most efficient location of a vent is at the end of fill of the part. As the plastic flows into the cavity, it begins to push the air out and continues to displace the air until the cavity is full of plastic. This means that the flow rate of the air leaving the mold is equal to the flow rate of the plastic entering the mold. If there was only one vent at the end of fill, the size of the vent would have to be equal to the size of the gate. But if this was the case, the plastic would easily get out of the mold through the vent during the pack and hold phase of the molding cycle. Because the vent depths have to be substantially smaller, the required amount of cross sectional vent area must be distributed across the mold. The area most accessible in the flow path of the plastic is at the parting line of the mold. Therefore, the parting line must be vented as much as possible. Once the cavity in the area of the parting line is filled, the plastic now pushes the air to other areas of the mold. Any ejector pins, core pins, cavity inserts, or cavity blocks in these areas should be used to help vent the air out of the mold. For each of these components, vents must be added based on the recommended vent depths. Ejector pins offer the most advantage, because they are self-cleaning. The residue from the gasses released through the ejector pin vents will get cleaned as the ejectors cycle back and forth during every cycle. Using ejector pins for venting does build up residue elsewhere in the ejector box and increases the frequency of maintenance. The vents on stationary components, such as the core pins, can get plugged easily and may need frequent cleaning. Sometimes a positive air blast during the ejection can help clean the vents. In recent years, porous steel has become increasingly popular. For example, in deep draw parts, where placing an ejector pin is not possible or not acceptable, the end of fill areas are made of porous steel inserts that allow pressurized air to escape from the cavity. These inserts do require frequent cleaning and therefore the molds are typically built such that these inserts can be taken out of the mold while the mold is still in the machine. The inserts are removed, cleaned, and put back into the mold while it is still bolted into the molding machine. Using porous steels does not allow for mirror finishes or high polish surfaces.

One area of the mold that is often not given much consideration with respect to venting is the runner. If the runner is not vented, the complete volume of air in the sprue and runner enters the cavity through the gate of the part. This air burdens the cavity vents with additional work. Runners must therefore be vented as much as possible and even up to a depth where some flash can be seen. It does not really matter if the runners exhibit some flash because the molded parts are of primary interest. The sprue puller pin must also be vented. Often, at the end of the screw recovery, the screw is sucked back without rotation to relieve any pressure on the melt and prevent it from drooling from the nozzle tip. This causes air to be sucked into the machine through the nozzle. Because the air is present in the heated nozzle with some of the plastic, there can be some build-up of gasses in the time after screw recovery and before injection. During the start of injection, this air and gas is injected into the mold. This is another reason that the runners must be vented. Figure 11.11 shows a mold with possible vent locations on the parting line.

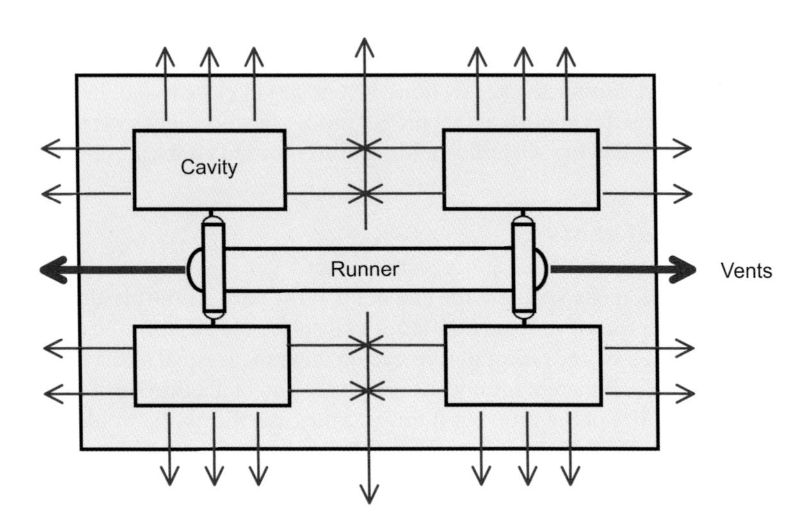

Figure 11.11 Possible location of vents on the parting line

11.2.4 Forced Venting or Vacuum Venting

In every mold there will be corners where air will get trapped. Because the volume of the air may not be large, defects are not easily seen. Close examination will show a rounded corner rather than a crisp corner. In other molded parts, shorts may be easily seen with the naked eye. Placing ejector pins in these areas to provide venting may not always be possible. If the corner detail is important to the function of the part, the only solution to eliminate the defect is to install vacuum in the mold. The vacuum evacuates all air from the system before injection, so that all cavity details fill with plastic, thus producing an acceptable part. Molds that need to be fitted with vacuum have parting line vents in the cavity blocks that are relieved into an outlet to which the vacuum line is attached. There is a seal around the cavity inserts to prevent outside air from being sucked in. This seal is formed when the mold is closed. Molds with slides and other contours are not good candidates for adding vacuum vents, because their mold faces are not always flat. Adding the seal may require additional real estate, increasing the size of the mold. The ejector pins may also need seals to hold the vacuum. Cold runner molds have an open sprue and will leak the vacuum. Although vacuum can be used successfully on cold runner molds (by slight modification of the process), valve gate molds and hot runner molds are better candidates for vacuum assist. In many cases, vacuum is added after several attempts of adding conventional vents have failed to produce acceptable parts. By then, all the ejector pins and/or core pins are vented and the parting lines, cavity blocks, and slides have been relived to the atmosphere. Additional work may now be required to fix these so-called leaks in the molds. Venting must therefore be considered at the mold design stage and the product designer must clearly specify the requirements on the part print. Figure 11.12 shows a mold configured for of vacuum venting.

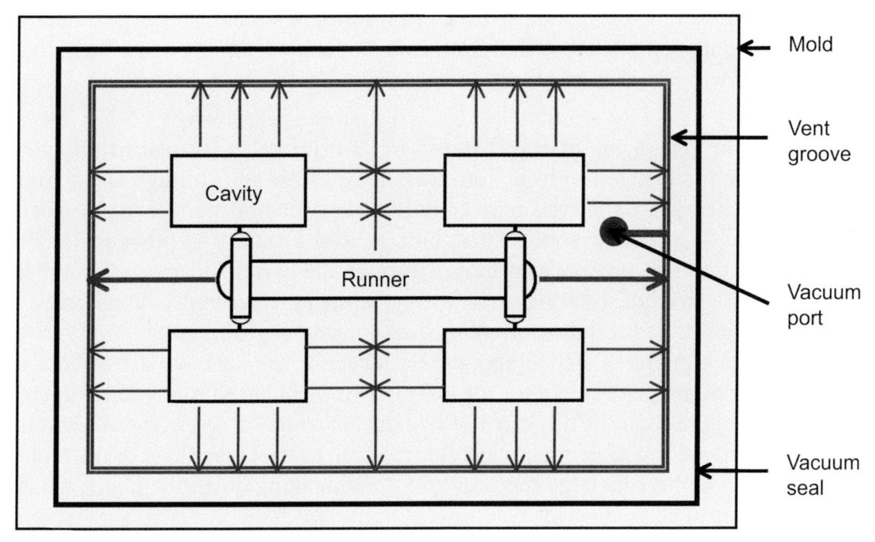

Figure 11.12 Venting using vacuum

11.3 Regrind

In injection molding manufacturing processes, runners that fall out of the mold or parts that are found defective can be ground up for reuse. This material is called regrind. Regrind material is generated from plastic that comes with at least one heat history cycle at the processing facility. Materials supplied by the resin supplier are classified as virgin materials, even though they may have gone through a melt processing process for compounding of additives such as fillers or colors. Regrind material is traded either directly or through third parties between companies that cannot use regrind material, such as companies manufacturing medical applications and those companies that produce certain consumer products, such as garden equipment or trash carrying equipment, who can use some amount of regrind and still manufacture products within their product requirements. Park benches are an example for a product where 100 % regrind material is used.

11.3.1 Effect of the Molding Process on the Part Properties

The molding process has been redefined as the collection of all the processes the plastic is subjected to from the moment it enters the manufacturing facility as loose pellets to the point when it leaves the facility as a finished molded product. The process can be broken down into the drying, melting, mold filling, cooling, and the final packaging of the part. Each phase of the process can affect the plastic, the additives in the plastic, and the fillers. During the drying process there is no breakdown in the molecular weight of the plastic. However, as discussed in Chapter 4, exposure of the material to excessive drying times can cause a loss or breakdown of the additives. Drying times in some resins are cumulative. If the lost additives happen to be those that contribute to the heat stability of the melt, then the plastic molecules will break down during melt processing. Fillers are not affected by excessive drying times.

During the melting phase, the plastic is transformed from pellet to melt in the injection barrel. The plastic is subjected to heat from the heater bands and through shear from the rotating screw.. The plastic material may contain fibrous or non-fibrous fillers. There are other additives, such as plasticizers, heat stabilizers, and a variety of other materials that are compounded into the material as it moves through the barrel. The heat from the heater bands and the shear from the rotating screw will cause the polymer molecules to break down and it is the extent of this breakdown that must be considered. Processing the plastic at the high end of the recommended melt temperature can be increasingly more detrimental to the molecular weight of the plastic because the plastic is now subjected to higher temperatures where the allowable residence time may not be large. Having the screw speed set at a high value can also increase the shear rate, again contributing to the breakdown of the molecules. High screw speeds can create large mechanical forces breaking down the fibers, while non-fibrous fillers may not be as affected.

During the mold filling phase, the material passes through the nozzle, the sprue, the runners, the gates, and into the cavity and finally reaches the end of fill. During this filling phase, the material is subjected to high shear forces, causing a breakdown in the plastic molecules. The fillers may not be impacted, unless the gate sizes are small. Shear is higher in smaller diameter runners and therefore smaller runners cause larger drops in molecular weight. Large tab gates offer the least amount of molecular weight decrease through the gates. The part geometry can also play a role in the degradation of the plastic. Thinner cross sections increase the shear and therefore increase the chance of plastic degradation. Insufficient venting in the mold can cause a dieseling effect and burn the plastic which in turn lowers the molecular weight. This may be difficult to observe on dark color parts.

The discussion above shows that each phase of the process has an effect on the molecular weight of the plastic and the integrity of the fillers. If any one of these processes is not controlled properly, it is easy to degrade the plastic. In many cases, plastic degradation or filler breakdown may not be cosmetically noticeable, making it easy to assume that the process was acceptable.

The generation number for a quantity of regrind is defined as the number of times the regrind has been generated from the same batch of plastic. For example, if a runner was ground up to be used back into the process, it is considered as first generation regrind. When the regrind is reprocessed back into the machine and regrind is generated from the molded runner, then this regrind is called second generation regrind. With each progressive generation, there is a deterioration of the plastic properties.

When regrinding, the runner or the parts are reduced to a size close to the virgin plastic pellets with the help of a granulator (grinder). Because of the random mechanical breakdown of the plastic, the particle size distribution ranges from fines to coarse particles. Low rpm grinders are best suited to achieve a narrow particle size distribution. A large amount of fines can cause two problems: First, due to a lack of mass, the fines melt immediately and stick to the feed section of the screw, causing inconsistent screw recovery. Second, these fines tend to degrade much faster, causing defects in the part. Fines must therefore be removed before they enter the feed throat. This is done with fine separators also called cyclone separators. In addition, regrind pellet size must be close to the virgin pellet size to achieve consistent mixing and feeding of material that contains regrind.

11.3.2 Using Regrind

It is clear that using regrind can cause a varying degree of loss in material properties. Therefore, the amount of regrind that can be used in a product will depend on how this loss affects the performance of the final product. It is impossible to provide general hard and fast rules or equations to calculate the acceptable amount of regrind that can be used in a part. The only reliable method is to experiment with varying percentages of regrind and different generations of regrind. A good starting point may be to run the product with 100 % first generation regrind and perform functional testing on the parts. Surprisingly, on some products this may be completely acceptable. If the initial process was carefully controlled and if the molding process was developed using scientific molding principles with large process windows, the material properties may stay well within the required specifications. If the 100 % product is unacceptable, various ratios of virgin to regrind must be evaluated. When testing the product, the properties that are important to the product must be evaluated. Testing of the product in its final assembly and service environment is critical. Lab results from tensile or impact testing can only provide a comparison of the property, but product testing results will provide the actual value. The testing procedures must be standardized and should be used in regular production. An acceptable part can also be used to create a standard which all products can be compared to or checked against to determine their acceptability.

11.3.3 Batch and Continuous Processes of Incorporating Regrind

There are two methods of incorporating regrind into the product. In the batch process, the regrind is generated offline, away from the molding machine. A predetermined amount of regrind is either blended with the virgin material and loaded into the machine hopper, or the regrind and virgin material are blended together at the molding machine hopper, using blending equipment. In this system, the regrind will have to go through the drying process together with the virgin plastic. Blending equipment based on weight (gravimetric blenders) is most reliable. Some systems rely on loading time of virgin plastic and regrind to achieve the right percentage. For example, to achieve an 80:20 blend the virgin plastic is fed into the system for 80 seconds and the regrind for 20 seconds. This system is not reliable because time does not equate to weight considering the possible differences in the bulk densities of the virgin plastic and regrind. If the plastic gaylord or container, which feeds the hopper, runs out of material, it may only be loading either the virgin or the possible worst case, only regrind. Gravimetric feeders look for a particular amount by weight of the virgin and regrind before delivering the mixture to the molding machine. If one of the components is not available, either the blending equipment or the molding machine will set off an alarm for lack of material. Blending equipment is also physically mixing virgin material and regrind before delivering it into the hopper of the molding machine. Systems that rely on time may end up having layers of virgin and regrind in the hopper because they alternately deliver each component for the specified length of time to the hopper of the molding machine.

In a batch process, the chances of contamination are very high and this is probably the main reason why regrind projects often are not successful. It is easy for two runners of the same color but of different materials to be thrown into the same grinder. Contamination can show up right away when the barrel is being purged for molding. If the contaminant melts at a relatively lower temperature but is now subjected to a higher temperature, it can easily start gassing or spurting. It may also show up as splay or other cosmetic defects. If the contaminant is a higher melting plastic, there is a possibility of it getting stuck in gates and hot tips or it may show up as unmelted pellets in the part if the gate sizes are large. In both cases, delamination, structural integrity or other issues can be the result and the part quality will suffer.

With a continuous process of regrind incorporation, the runners are picked up with the help of a sprue picker or robot and directly dumped into a granulator; the grinder grinds up the runner and delivers the regrind to the hopper immediately. The runners are not separated in a separate area to be mixed later, such as in the batch process. The continuous process is a cleaner way of incorporating the regrind with almost no chances of contamination. The disadvantage here is that some molecules have gone through the process multiple times because this is a continuous process and the runner is always recycled. It is impossible to figure out the generation of the regrind, because it is all one homogeneous mixture. The smaller the ratio of part to runner weight, the greater is the percentage of older generation regrind present in the system. Continuous processes are suitable for parts that are able to accept a large percentage of regrind and that have a larger ratio of part to runner weight. The smaller the runner, the lower is the amount of regrind and therefore the fewer are the chances of processing problems or product failure. If the runner is directly fed back to the feed throat, in most cases drying is not required, eliminating the risk of additive loss caused by the drying process.

11.3.4 Estimating the Amount of Regrind from different Generations

Consider a part to runner weight ratio of 80:20. Assuming the whole runner is incorporated back into the plastic, the percentage of each generation of regrind present at each pass of molding is shown in Table 11.1. The concept of generation at each pass may seem confusing. First generation regrind during pass four will be the regrind generated from the virgin material from pass three. It will not be regrind generated during pass one. The regrind generated during pass one becomes third generation regrind. Because part and runner are molded in the same shot, the percentage of regrind will be the same in the runner and in the part.

It is a common perception that as the number of passes increases, the material starts to deteriorate very fast. This is true in cases where the part to runner weight ratio is high; for example the runner weight is 75 grams and the part weight is 25 grams. In the above example, where the part to runner ratio is 80:20, the table shows that there is always 64 g or 80% of virgin material in the part, 12.8 grams of 1st generation regrind, 2.56 grams of 2nd generation regrind and so on. If a disciplined approach has been taken to dry the plastic and process the material, the properties of the plastic may well be within the required specifications. This will be true with filled or unfilled plastics and in most cases this should be accept-

Table 11.1 Percentage of 'g' Generation Regrind after 'p' Number of Passes for a Part to Runner Ratio of 80:20

Regrind generation number (g)	Pass (p)				
	1	2	3	4	5
0 (Virgin)	100	80.00	80.00	80.00	80.00
1	–	20.00	16.00	16.00	16.00
2	–	–	4.00	3.20	3.20
3	–	–	–	0.80	0.64
4	–	–	–	–	0.16
Total	**100**	**100**	**100**	**100**	**100**

Table 11.2 Weight of 'g' Generation of Regrind after 'p' Number of Passes in an 80 g part with a 20 g runner

Regrind generation (g)	Pass (p)				
	1	2	3	4	5
0 (Virgin)	80	64.00	64.00	64.00	64.00
1	–	16.00	12.80	12.80	12.80
2	–	–	3.20	2.56	2.56
3	–	–	–	0.64	0.51
4	–	–	–	–	0.13
Total	**80**	**80**	**80**	**80**	**80**

Table 11.3 Weight of 'g' Generation of Regrind after 'p' Number of Passes in a 20 g Runner of a 80 g Part

Regrind generation (g)	Pass (p)				
	1	2	3	4	5
0 (Virgin)	20.00	16.00	16.00	16.00	16.00
1	–	4.00	3.20	3.20	3.20
2	–	–	0.80	0.64	0.64
3	–	–	–	0.16	0.13
4	–	–	–	–	0.03
Total	**20**	**20**	**20**	**20**	**20**

able. If the part design was not robust, it is possible that the small decrease in the properties would result in the failure of the part.

The following formula can be used to estimate the percentage of regrind from each generation in a part:

(1) If $(p - g) < 1$, then $R = 0$,

(2) If $(p - g) = 1$, then $R = \left(\left(\frac{x}{100}\right)^g\right)100$

(3) If $(p - g) > 1$, then $R = \left(\left(\frac{x}{100}\right)^g\left(1 - \frac{x}{100}\right)\right)100$

with

x = percentage weight of the runner

g = generation of the regrind

p = molding pass number

R = percentage of regrind

Example

Part weight = 35 g, runner weight = 7 g

therefore, $x = \left(\frac{7}{7+35}\right) \times 100 = 16.67$

In the 4th pass ($p = 5$), the amount of 3rd generation regrind ($g = 3$) can be determined with $(p - g) = (5 - 3) = 2 > 1$ as

$R = ((16.67/100)^3 (1 - (16.67/100))100 = 0.39\ \%$

Regrind tables for various runner ratios are give in Appendix G.

11.3.5 Effect of Regrind on Processing

Regrind can have lower molecular weight, degraded additives and/or their byproducts, damaged or destroyed fillers, and contaminants. Depending on the extent of these changes it may be necessary to change the process to compensate for some of the changes in the outputs and make acceptable parts. A decrease in molecular weight will cause a drop in viscosity of the plastic, affecting all the outputs related to viscosity. Fill times and cushion values will be lower. If part of a processing aid or a viscosity reducer was lost, it is possible that the fill times and cushion values will be higher. If the filler was fibrous, the breakdown of the fiber length will cause the plastic to flow easier, resulting in shorter fill times and lower cushion values. Lower viscosity can also lead to an increase in plastic cavity pressure, although the pressure behind the screw remains unchanged. Excessive amounts of drecrease in molecular weight can cause the plastic to degrade easier, causing excessive gassing and plugging of the vents. Cosmetic defects such as splay can occur. These effects can be compensated for by changing some of the process parameters. For example, reducing melt temperature will increase the viscosity of the plastic or reduce the gassing that occurs. In all cases, it is important that the process be robust and more importantly be redeveloped based on the presence of regrind.

11.3.6 Closing Remarks

Using regrind is a good and effective way to save money and at the same time protecting the environment from waste plastic being dumped into landfills. However, many processors tend to switch back to virgin resin as soon as they detect a quality problem. Often the quality issue is fixed and the processors are hesitant to go back to using regrind because they identify the cause of the problem being the regrind. Although this could be true, the reason why the regrind was causing the problem must be investigated. Sometimes the regrind is not properly dried or there may be too many fines from the regrind process that could have caused the problem. A proper analysis of the amount of regrind and a procedure to incorporate it into the molding process are two very important steps that must be followed. The implementation plan can be pursued as far back as in the part design stage. A part designed to be molded from virgin plastic can be overdesigned to compensate for the loss of properties when regrind is used. Using regrind should be part of the production plan and as seen in the above 80:20 example, the part will always contain 92.8 % of virgin and first generation regrind resin and should therefore be able to function satisfactorily. Most companies that are not successful in implementing a regrind program fail because of lack of discipline and training of the shop floor personnel. For example, the operators who assist in grinding the runners need to be educated about the different types of plastics and that every clear runner in the molding facility is not necessarily the same material. Instituting a regrind program requires discipline and is a culture change.

Suggested Reading

1. Osswald, T.A., Turng , L., Gramann, P.J., *Injection Molding Handbook* (2007) Hanser, Munich
2. Beaumont, J.P., Nagel, R., Sherman R., *Successful Injection Molding* (2002) Hanser, Munich
3. Rosato, D.V., Rosato D.V, *Injection Molding Handbook* (2000) CBS, New Delhi, India

12 Related Technologies and Topics

This chapter will introduce a variety of technologies and techniques that can be applied to enhance the robustness of a process and/or to speed up the complete project from conceptualizing the part to its release as a molded product. Cavity pressure technology, which takes process monitoring and control to a completely new level, increasing product quality and plant efficiency many fold, will be introduced. Companies can build a knowledge base of their experiences to apply to future projects based on their past experiences. Concurrent engineering is an extension of the teamwork concept, keeping the whole team involved with the decisions and progress of the project.

12.1 Cavity Pressure Sensing Technology

When we discussed processing in the previous chapters, everything has all been related to the molding machine. For example, process optimization dealt with optimizing the processing parameters, such as injection speed and holding pressure. However, the part is finally made inside the mold, and therefore knowledge of what happens inside the mold can give the most valuable information about the quality of the part. The molten plastic will follow the specific volume–temperature graph (Fig. 2.10) that dictates the part quality. Tracing this information for every shot will reflect the quality of the part. Needless to say, if every shot follows the same curve each time, all shot will be identical. Although it is not easy to output a specific volume–temperature graph for the mold, there are other indirect methods. Placing a pressure transducer inside a mold can provide information about the melt pressure in the mold and as the melt starts to cool, the pressure decrease can be plotted. Temperature transducers inside the mold will provide information on the temperature of the plastic. Although the "transducer" is technically the right term, the term sensor is most commonly used and we shall adopt that terminology in this text.

12.1.1 Sensors and Output graphs

Figure 12.1 shows the placement of a 'Near the Gate Sensor' and an 'End of Fill Sensor' in a mold. A typical cavity pressure graph is shown in Figure 12.2. The graph also identifies the injection, holding, and cooling phases. There are three pressure traces in the figure that will be described in the following:

Hydraulic Pressure Curve: To measure hydraulic pressure, the sensor is placed in the hydraulic line. Since the injection phase is a fast dynamic phase, the pressure increases very rapidly until the start of the holding phase is reached. Once the switch-over from injection to holding takes place, the pressure drops to the holding pressure value and stays constant until the end of the holding time. After the holding time is complete, the hydraulic pressure drops to

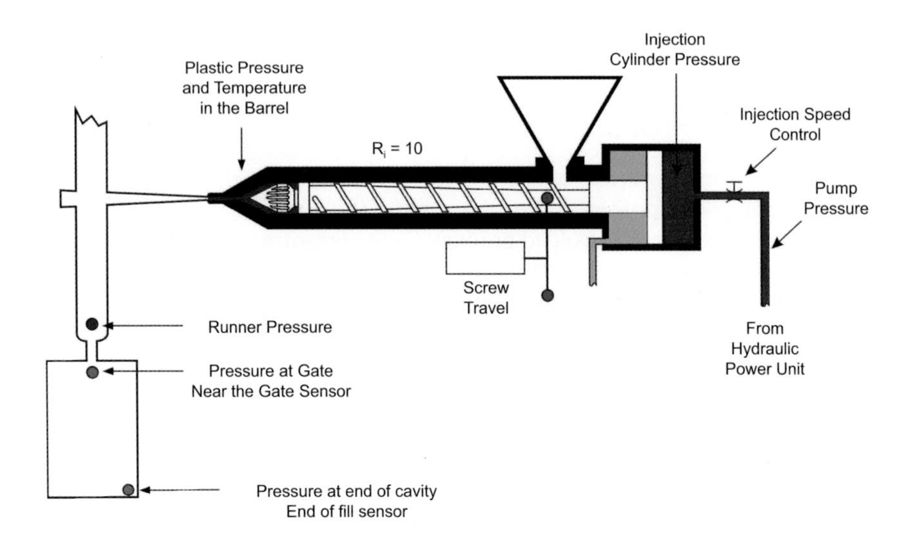

Figure 12.1 Schematic of placement of sensors inside a mold (Courtesy: RJG Inc.)

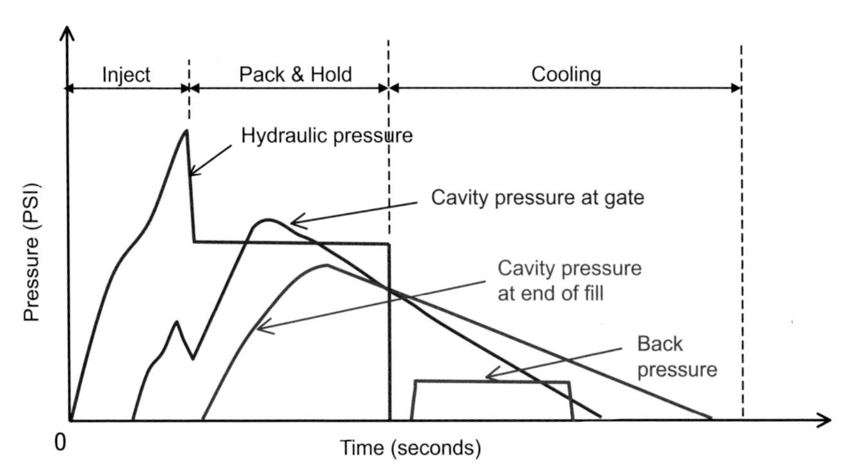

Figure 12.2 Typical representation of a cavity pressure graph

zero. If the screw rotation starts immediately, the back pressure shows up right after the end of the holding phase. If there is a delay, the trace picks back up indicating the back pressure for the screw recovery time.

Cavity Pressure Curve Near the Gate: As the plastic begins to enter the mold and the cavity, it is in contact with the sensor placed just behind the gate. The pressure increases until the injection phase is complete, but does not drop down like the hydraulic pressure, because the cavity is pressurized and the gate is frozen or closed. The decrease in melt pressure is caused by the shrinkage taking place inside the mold and it causes the plastic to move away from the mold wall and the sensor. The start of the curve has a delay which is equal to the time the plastic takes to reach the sensor from the start of injection.

The Cavity Pressure Curve at the End of Fill: This curve looks similar to the curve described above, the only difference being the starting point of the curve. Here, the starting point represents the time the plastic requires to reach the end of fill. If the process is decoupled (see Section 7.5.4), this time is almost equal to the injection fill time.

12.1.2 Types and Classification of Pressure Sensors

There are various types of sensors used in plastics processing. Their classification is based on the type of technology used to collect information.

Strain Gage Sensors: The underlying principle of a strain gage is based on the Wheatstone bridge that has a network of resistance elements. The electric current flowing through this network is measured. When any external force is applied to the strain gage, the amount of current changes. This change in the amount of the current is proportional to the applied force and hence the amount of force can be determined.

Quartz-Based Sensors (Piezoelectric Sensors): Some materials, such as quartz, generate an electrical potential when an external stress is applied. The electrical potential is directly proportional to the applied force. A quartz-based sensor can therefore measure the force applied on the face of the sensor.

Another type of sensor classification is based on the location of the sensor in the mold.

Direct Sensor: These are also called flush mount sensors, see Fig. 12.3. Here, the sensor face forms part of the cavity wall. The sensor is fitted such that it comes in contact with the molten plastic and measures the pressure. Such sensors are not commonly used because it

Figure 12.3a Flush mount sensor (Courtesy: RJG Inc.)

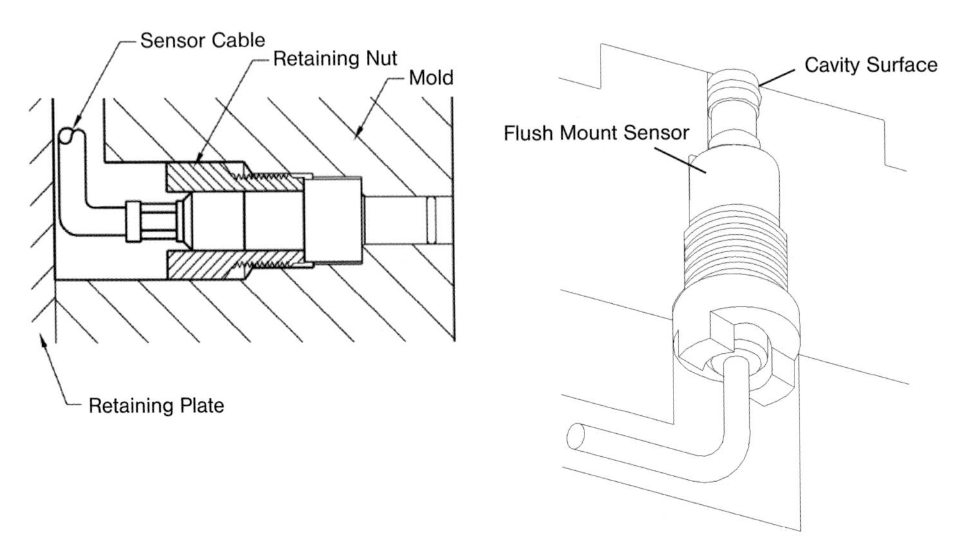

Figure 12.3b Flush mount sensor (Courtesy: RJG Inc.)

is not always possible to get the sensor inside the mold cavity. It can also be a risky task to work inside the cavity to fit the sensor. The cavity wall would need rework if damaged. Some molds are subject to high temperatures and this could also pose a problem because pressure sensors may not withstand high temperatures.

Indirect Sensors: When a sensor is placed underneath an ejector pin or a core pin, it is called an indirect sensor. The force is applied on the face of the pin and transferred to the base of the pin where the sensor is located, see Fig. 12.4. These sensors are more common because there is room to work with inside the ejector box where the sensor has to be fitted.

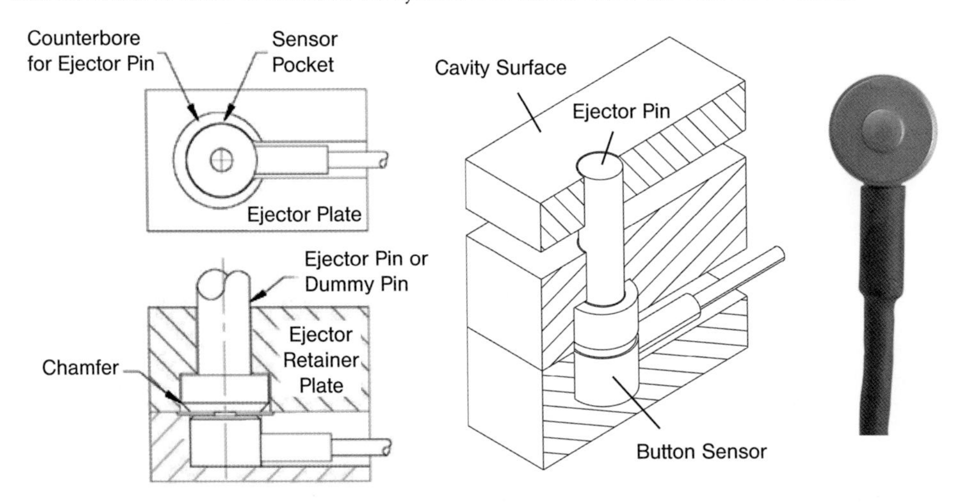

Figure 12.4 Button sensor behind an ejector pin (Courtesy: RJG)

Yet another type of classification of the sensors is based on the function the sensor will perform in the mold.

Control Sensors help control the process (which will be described in the following).

Monitor Sensors are used for monitoring the process. Process alarm limits can be set based on the information acquired from this sensor.

12.1.3 Use of Information from the Pressure Graphs

The information obtained from the pressure graphs is valuable in a number of ways.

- The pressure graphs that are traced provide information about the cavity pressure and therefore the part quality based on the specific volume–temperature relationship.
- Inconsistencies from shot to shot can be observed. If the pressure traces do not repeat, the part quality is not repeated, see Fig. 12.5. It can be observed that the hydraulic curve overlays itself every shot, proving the hydraulic pressure is consistent. However, the plastic pressure curves are not being duplicated.. If there was no cavity pressure hookup, one would automatically assume that the process is consistent and that the part quality should also be consistent from shot to shot. A leaking check ring can cause such a variation and can only be noticed with a cavity pressure sensor.
- If there are sensors present in each cavity, the quality of each cavity can be compared, see Fig. 12.6. In this example, there are four cavities and the traces for each cavity are different. Based on the specific volume–temperature curve, each cavity will produce different parts. Although there is consistency within a cavity, there is cavity-to-cavity inconsistency.
- The cavity pressure graph can also provide gate seal information. If the gate is not sealed, the plastic will leak out of the cavity, causing a sudden drop in cavity pressure. The curve

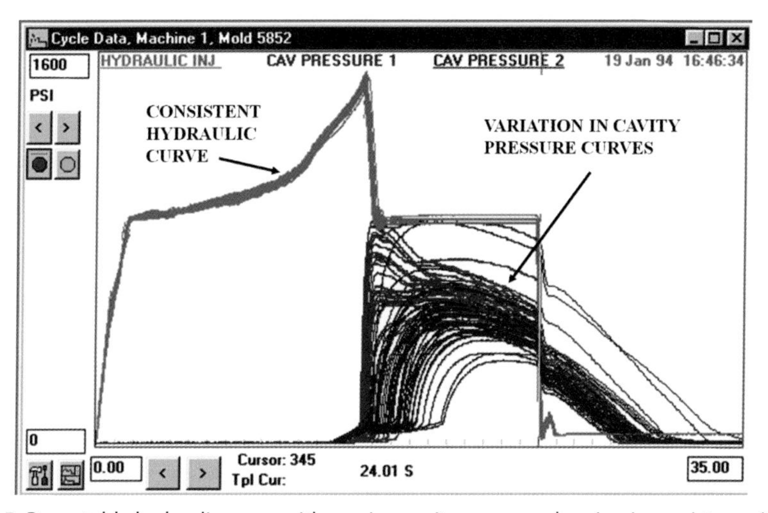

Figure 12.5 Repeatable hydraulic curve with varying cavity pressure showing inconsistency in part quality (Courtesy: RJG Inc.)

will therefore show a sudden change in its slope. A frozen gate will show as a gradual drop in cavity pressure rather than a sudden change in slope, see Fig. 12. 7.

- If the sensor is at the end of fill, a short shot will result in a 'zero' value for the pressure. This can be used to detect short shots and contain the parts. Usually, a signal is sent to a robot to reject the parts or sent to a conveyor to reverse and reject the parts.

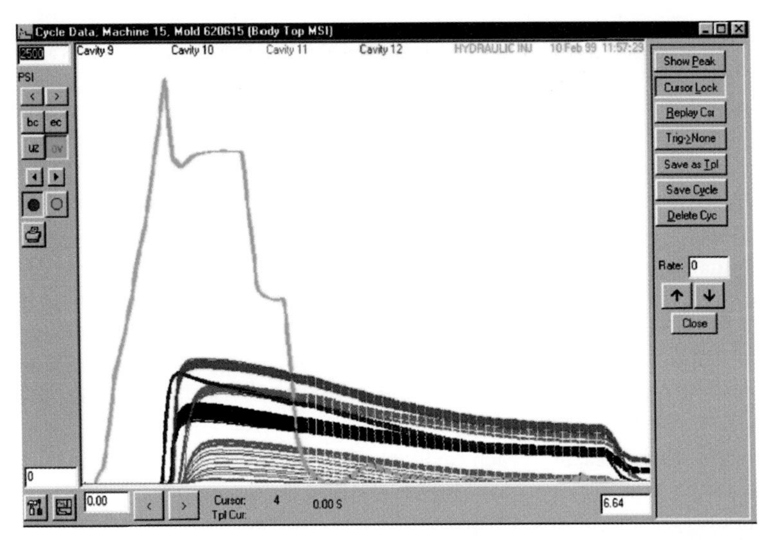

Figure 12.6 Cavity pressure variations between cavities showing cavity-to-cavity inconsistency, note that cavity 11 has the greatest variation, ranging from well-packed to nearly short (Courtesy: RJG Inc.)

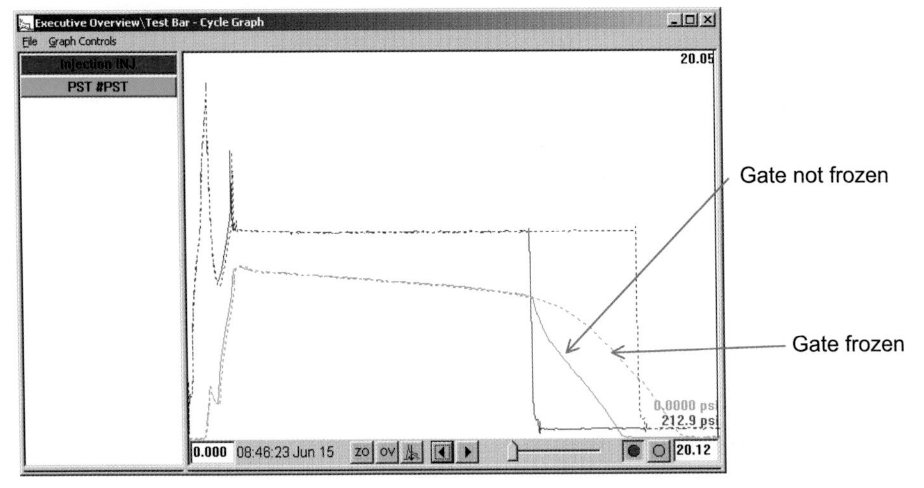

Figure 12.7 Cavity pressure curve showing the status of gate freeze (Courtesy: RJG Inc.)

12.1.4 Controlling the Process with Cavity Pressure Sensors

A decoupled process is the most efficient and consistent method of molding. The switch-over from the injection phase to the holding phase can be initiated by monitoring one of the following parameters: hydraulic pressure, time, or screw position. The most common and consistent method used is the screw position. When the screw reaches a preset position, it transfers to the holding phase. The screw position controls the percentage of the volume filled and, when using decoupled molding (see Section 7.5.4), this percentage should be between 95 and 98 %. In response to the injection phase there is an increase in the cavity pressure and then a slow decrease. Figures 12.8 through 12.11 show the various cycle integrals.

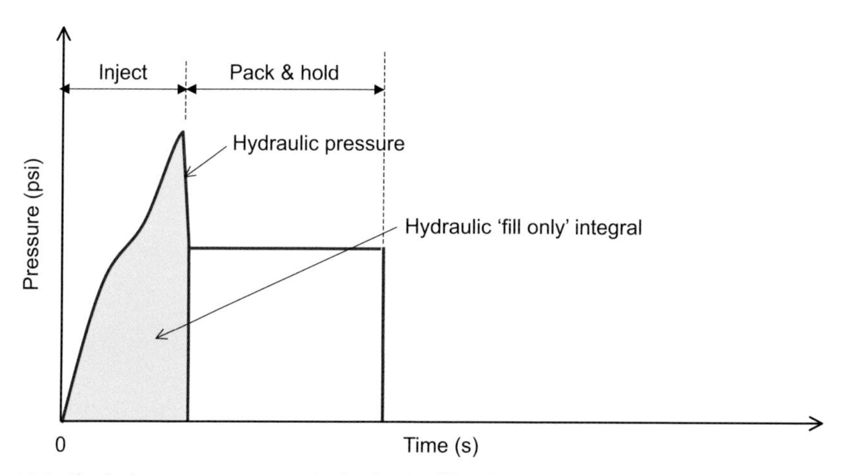

Figure 12.8 Shaded area representing the hydraulic 'fill only' integral

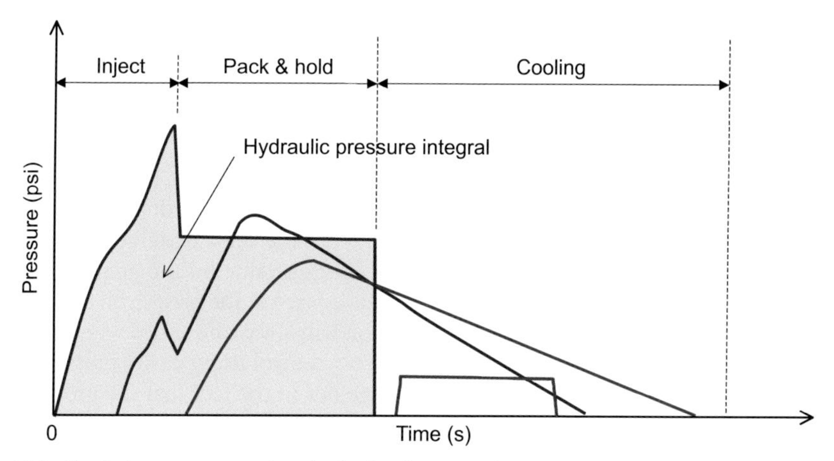

Figure 12.9 Shaded area representing the hydraulic integral

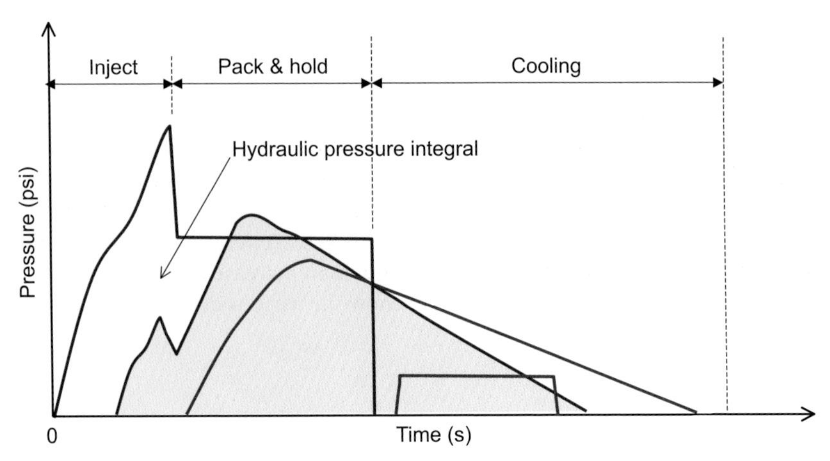

Figure 12.10 Shaded area representing the post-gate cavity pressure integral

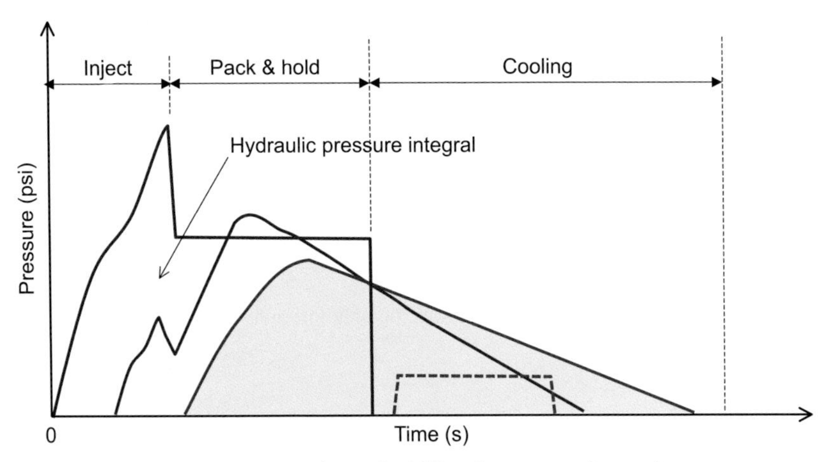

Figure 12.11 Shaded area representing the end-of-fill cavity pressure integral

Peak pressure, time to the peak pressure, and the area under the curve before the peak pressure (shaded area) represent the injection phase. Therefore, if this is duplicated every shot, the fill is duplicated every shot. If the information from the sensor can be fed back to the machine and if the switch-over is now controlled using a value on the graph (usually the peak pressure), the switch-over for every shot will take place at the precise fill amount of the cavity, resulting in shot-to-shot consistency. Injection will always be ended when the desired pressure is reached in the cavity. This is called process control using cavity pressure sensors. Other than process consistency, the other advantage lies in the fact that the process is now machine-independent. If the same curve is repeated on another machine, the part quality will be the same. Controlling the process using pressure sensors leads to the most consistent process.

12.1.5 Sensor Locations

Sensor locations are important because the location will dictate the amount and quality of the information it will acquire. Preferably, sensors to control the process must be located near the start of the fill (in the first third of the part) and the sensors to monitor the process should be near the end of the fill, in the last third of the part. For extremely small parts, a sensor can be located in the runner. The compromise here is that real gate seal information is not available because it can be obtained only from the pressure traces in the cavity. Table 12.1 shows the benefits of certain types of sensors and their locations (direct or indirect).

Table 12.1 Comparative Benefits of Types of Sensors and Technology (Courtesy: RJG Inc.)

	Types of sensors and technology			
	Flush mount (direct)	**Under the pin (indirect)**	**Piezo-electric**	**Strain gauge**
Accuracy of data	Yes	Yes	Yes	Yes
Ease of calibration / configuration	Yes			Yes
Location flexibility	Yes			
Lower installation cost		Yes		
Easier replacement		Yes		
Removable cables			Yes	
Limited mold space	Yes		Yes*	
High temperature			Yes	
Easy troubleshooting		Yes**		Yes
Use with long cycles				Yes
Lower costs				Yes

* Smaller connector

** Easier removal

12.2 Building a Knowledge Base

Many molding companies try to find and dominate a niche in the market. For example, a molder may be known to be an expert in molding gears with high precision while another may be an expert in two-shot molding. Companies reach these levels of expertise having perfected the technologies over a number of years. The collected knowledge comes from a pool of experience, trial and errors, and applied science. This knowledge and solutions to the problems should be recorded to make future projects efficient during project development, the release to production process, and finally in production. For example, consider a company that molds connectors of various types, primarily from nylons and polyesters. Over the years they may have built hundreds of tools for similar types of connectors. There may be a commonality in the shape of the connectors, the gate locations and sizes, cooling channel locations, wall thicknesses, and so on. If the scientific processing principles are followed, the processing parameters would also end up being very similar. For example, the mold and melt temperatures or the plastic pressures in the cavities can be very similar for parts being molded from the same materials. Similar processing parameters will lead to similar shrinkage values. This can be verified by taking the steel dimension and comparing it with the part dimension. If the cavity steel for a number of molds and their respective molded parts are measured, a database can be built and a simple formula or a trend graph can be developed. When a new similar project comes in, this database can then provide a better estimate of the shrink values that can be used when designing the mold. This technique can be very useful when designing components that are difficult to make and when mold dimensions cannot be kept steel-safe. A threaded insert is a good example of this type of component. A 16-cavity threaded core insert would be very expensive to rework, if the final thread size on the molded part was not within specifications and not within the required capability. Here, prior knowledge of a similar mold would come in very useful. If the plastic flow is similar in an existing mold compared to a new mold, the molecular orientations can be similar, leading to similar shrinkage values. If the process using the existing mold was robust, using similar parameters will lead to a robust process in the new mold. Again, the part quality becomes predict-

Summary of Mold Dimension, Part Dimension and Process Parameters for Part No.: 84736-2

Part Description	: Cover Connector	*Melt Temperature*	: 480 deg F
Material ID	: PBT	*Mold Temperature*	: 160 deg F
Grade	: Valox 420 (Sabic IP)	*Fill Time*	: 0.98 sec
% Filler	: 30% glass	*Cavity Pressure - Post Gate (Max)*	: 8500 psi
Runner Size feeding the part	: 0.3175 inch	*Machine Shot Size*	: 150 gms
Type of Gate	: Sub D-gate	*Barrel Usage*	: 65.75 %
Number of Gates	: 1	*Machine Tonnage*	: 175 tons
Gate Size	: 0.060 inches		
Number of Cavities	: 4		

Critical Dimensions	*Steel*	*Part*	*% Shrinkage*
D1 - Length	2.540	2.502	1.5
D2 - Width	1.740	1.705	2
D3 - Shroud Height (@ C)	0.507	0.497	2
D4 - Shroud Height (@ B)	0.507	0.492	3

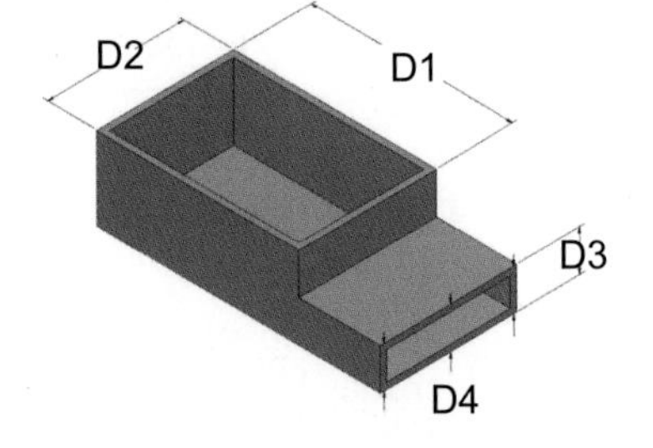

Notes:
Venting affects the fill and the dimensions of the shroud. Build shroud for easy access of vents. Vents must be cleaned at the start of every shift.

Figure 12.12 Knowledge worksheet

able. In referring to the threaded part mentioned earlier, determining the size of the cores can be done with a higher confidence level. Although shrinkage can be estimated with the help of simulation programs, the data acquired by modeling is rarely accurate, because of the basic assumptions that go into building the software algorithms. For example, a software program will assume perfect venting and perfect packing of the parts, although in practical molding this is never the case. Building a knowledge database for similar products eliminates the need for assumptions and at the same time takes into account all common factors, making the estimation more reliable. It is important to keep the steel safe and to start the mold dimensional modifications only after the first mold trial where a robust process was developed.

Building a knowledge base helps taking the guess work out and reducing the number of iterations to the final product release to manufacturing. This saves time and money, making the overall process efficient. An example of a summary sheet is shown in Fig. 12.12. The data can also be used to build a second mold for an identical part, especially when in the first mold some of the required part dimensions were unattainable, possibly because of the cost or time of modification or of making new mold components. In such cases, the mold design can be reverse-engineered and the parts in the new mold can be molded within the required specifications.

12.3 Concurrent Engineering in Injection Molding

A molding company usually receives a new project request through the sales department. The sales department then asks the engineering department about the feasibility of the molding. Once the job is accepted, the tooling department gets involved. They design the mold and help building the mold through the mold maker. The mold is then mounted on the machine and the expectation is that the process engineer can mold acceptable parts in the first attempt. Unfortunately, molding the parts successfully in the first attempt is not common, because the process, although having followed a logical sequence, did not involve all departments in the appropriate sequence of the decision making process, leading to certain failures. Many mistakes are often discovered during the first trial of the mold. 'They should have done it this way' or 'The parts will never fall off the mold' are typical comments during the first trial of the mold. This could have been avoided, if all departments were made aware of the project and its requirements before the mold was built. 'Over the wall engineering' is a term used to describe the project activity where each department is disconnected from each other. Each department performs the activities requested by the previous one and passes the project on to the next (Fig. 12.13.)

'Concurrent Engineering' involves representatives from every department in the project at the earliest stage (Fig. 12.14). This should be followed by subsequent meetings, where the project details and any changes should be reviewed by the team to evaluate the impact a change may have on their own individual departments. Therefore, when the product designer mentions that the material is, for example, a long fiber thermoplastic (LFT), the process engineer knows, he will need a different screw and nozzle tip, or the scheduler knows that he will need to schedule time on the one specific machine in the shop with the required special screw. Raising a red flag at this stage saves time and money and delivers the product on

Figure 12.13 Over the wall engineering – departments are disconnected from each other

time. Although conventional 'Over the Wall' approaches are commonplace in most companies, disengaging each department is a common cause of failures, inefficiencies, and delayed launch of the product.

The concept is further explained in the matrix in Table 12.2. The first column is a typical list of departments involved in the design and manufacture of a product. The first row describes the activities involved. A 'Yes' at the intersection of a row and a column means that the particular activity has an impact on that department. For example, for the process engineer, the mold design is very important, while for the quality department, the mold design has no direct impact. Please note that this is a universal matrix and there will be exceptions. This matrix should be used as a guideline to set up individual company matrices.

Figure 12.14 Concurrent engineering

Table 12.2 Suggested Personnel – Activity Chart for Implementing Concurrent Engineering

Job function	Activity						
	Product design	Material selection	Mold design	Mold construction	Machine selection	Process development	Quality requirements
Product designer	Yes	Yes	No	No	No	Yes	Yes
Tooling engineer	Yes	No	Yes	Yes	No	Yes	Yes
Mold maker	No	No	Yes	Yes	No	Yes	No
Material supplier	Yes	Yes	No	No	Yes	Yes	No
Process engineer	Yes	Yes	Yes	No	Yes	Yes	Yes
Quality engineer	No	Yes	No	No	No	Yes	Yes
Sales team	No	Yes	No	Yes	No	No	Yes

12.3.1 The Product Designer

Conventional Involvement: Product design is where the concept first takes shape through a CAD model or a prototype. At this time, there is a general idea of what the requirements for the plastic material are and therefore a selection may or may not have taken place. The machine, the mold design, the molding process, or any of the other factors are of no concern to the designer, who is focused on a functional concept.

Required Involvement: A product designer should understand the manufacturing process, especially the processing part of it. Design for manufacturability principles must be implemented. For example, the designer must be aware of the fact that a thick section in a part must be cored out to reduce sink or sufficient draft must be provided on the part to release the part from the mold. An explanation of the function of the part to the molder and his team would be beneficial to the molder. This way, the process engineer knows the particular material he needs to process or can raise a flag when tolerances are impractical for a specific material.

12.3.2 The Tooling Engineer

Once the product designer presented an acceptable design, the tooling engineer gets involved in the design and the economics of building the mold. Of all job functions, the tooling engineer is probably the one who is most involved in the project because his direct involvement is required at the various stages.

Conventional Involvement: The tooling engineer is usually the liaison between the mold maker and the molder. His job is to deliver the mold to the molder and fix any issues the molder discovers during processing. Any design changes to the parts – features and/or dimensions – need to be reflected in the mold and the tooling engineer is responsible for these changes.

Concurrent Requirement: A tooling engineer must understand the concepts of developing a robust process. The techniques and the benefits of Scientific Processing should be familiar to him. The tools must be qualified using these techniques and changes to the mold must be made to support a robust process. For example, identifying a process window is extremely important. The tooling engineer must closely work with the process engineer, the quality engineer, and the mold maker.

12.3.3 The Mold Designer and Mold Maker

Most mold makers work with their own mold designers. They are given the part model and design the mold accordingly.

Conventional Involvement: The mold designer receives the design from the part designer, designs the mold for the chosen number of cavities and the mold maker fabricates the mold. He knows the type of material that will be used for molding the part, but sometimes the details are not made available to him.

Concurrent Involvement: The mold designer and the mold maker must get in touch with the material supplier to get information before designing the runners, gates, and vents. For example, it makes a big difference whether a 30 % glass filled material or a 30 % long fiber glass filled material is used. Apart from this, the mold design must be reviewed between the part designer, the mold designer, the mold maker and the process engineer. When the first shots are produced the mold maker must be present to see how the mold functions. The mold maker must also understand the concepts of Scientific Processing and why a process window is important.

12.3.4 The Material Supplier

In most cases, the material suppliers sell the resin but do not get involved, unless there is a processing issue. Since they are the suppliers of the materials they have all the information such as data sheets, processing information, mold design requirements, part design requirements and are therefore the best source of information.

Conventional Involvement: They provide material specs to the designer, help with material selection, and sell the resin.

Concurrent Involvement: Material suppliers will be able to review gate types, gate sizes, and vent sizes for any given material. For example, sub-gates are popularly and commonly used because they automatically separate the part from the runner; however, they should not be used for highly fiber-filled materials. In case a material is molded for the first time in a particular facility, the molder must invite the material supplier. The material supplier can make sure that the material is processed properly for optimum properties and cycle time. It also helps to involve the material supplier at the mold design stage, again in particular if this is a new material.

12.3.5 The Process Engineer

Unfortunately, often the process engineer is the least involved member of the whole team when in fact he is the most important one. The process engineer is the person who has to deliver the final molded product. Project managers are anxious to see the first molded part in their hand and therefore are usually waiting at the molding machine on the day of the trial. Every other stage of the project, for example mold build, typically has an end date, but the process engineer is under pressure to deliver the parts hours after the mold has arrived into the molding facility and minutes after the mold is mounted in the molding machine. Most likely, the project manager has promised the customer parts by next day delivery or may have also invited the customer to be present for the trial. This puts tremendous pressure on the process engineer who is expected to mold acceptable parts. However, in many cases the process engineer has to take the blame for all the factors that were not taken into consideration and all the mistakes and miscommunications during the entire project if the parts were not acceptable.

Conventional Involvement: In some organizations the process engineer sees the mold or knows about the project on the day it is scheduled and in some cases they are involved at the mold design stage.

Concurrent Requirement: The process engineer must be involved at every stage of the project. Based on his molding experience, he can give a number of suggestions to improve the moldability and the molding process of the part. Process engineers are better judges of features such as vents, gate locations and so on. Mold designers tend to place gates in locations convenient for mold making which are not always the best locations regarding processing. Even non-technical requirements, such as the orientation of the mold, should be reviewed by the process engineer. The choice of the machine must be left to the process engineer. Again, based on his calculations of tonnage, % shot size used, and residence time, he can suggest the best machine for the job at hand.

12.3.6 The Quality Engineer

Typically, the quality engineer is only a 'measurement guy' and often not involved appropriately. The quality engineer past experience with similar parts or plastics can prove valuable. He usually has knowledge about shrinkages and appropriate tolerances during production.

Conventional Involvement: He is one of the invitees to the product release meetings and is usually assigned the task to obtain drawings and work on methods for measuring the final molded parts.

Concurrent Requirement: The drawings must be discussed with the product designer. If an stereolithography model of the part or a prototype model is available, it should be used to detail the inspection plan. Sometimes fixtures are required and these can be planned in advance. Initial Gage Reproducibility & Repeatability (Gage R&R) studies can even be performed on prototype parts. Any impractical tolerances can be mentioned to the designer in advance.

12.3.7 The Sales Team at the Molder

The sales team is usually the one that introduces the product designer to the molder. They are looking to increase the sales of the molder and therefore try to acquire as much business as possible. However, they must understand the technical aspects of the molding in order to get the right set of customers. They must evaluate if an incoming project is suitable to and compatible with the capabilities of the molding operation. If product and molder are not the right match, this could have a negative impact not only on the program but also on the relations between the two parties. All future work between the two parties can get jeopardized.

Conventional Involvement: The sales force brings in a job mainly based on machine tonnage. Then there are other factors that they consider, such as any special requirements. Special requirements include clean rooms, in-mold decoration, or insert molding.

Concurrent involvement: The sales team must understand the capabilities of the molder. This also includes the strength and weaknesses of each department in the organization. If not, he is setting up the company for failure. The rules of machine selection for a particular job must be clearly understood. Tonnage, shot sizes, percentage of shot size used, residence times and other parameters must be known before accepting a job.

12.3.8 Mandatory for All Departments

The final molded part is the result of the efforts of all departments involved. In traditional 'Over the Wall' practices, every department performed their list of things to do and passed the project on to the next department. However, the practice fails to address the issues the receiving departments could have as a result of an action from a department previously involved. Understanding the needs of each department and moreover a 'why these needs' makes the jobs easier, efficient, and ensures on-time delivery of a part that meets the specifications. Every department must understand cross-functional duties and have a basic understanding of each other's job function.

For example, everyone must be trained scientific processing methods and understand why it is important to have a good process window. If the process engineer sends a mold back to the mold maker or the tooling engineer because he does not have a sufficient process window, he has a very valid reason. He can probably mold ten good parts, but that does not mean that he can make half a million good parts, because the process will not be capable. Knowledge of process capability must be commonplace in every department of the molding facility.

12.3.9 Implementing Concurrent Engineering

Concurrent engineering is probably the easiest concept to implement, because all it takes is to get all the involved departments in one room or on a conference call to discuss the project. Table 12.2 provided a simple matrix of typical job functions and activities involved. Set up a meeting between the representatives of each department for the various phases of the project. The following must be taken into account when following the table.

1. The order of certain decisions does not necessarily reflect the order in which certain activities are performed. For example, a machine must be selected at the quoting stage to make sure the molder has the machine for the particular job and not when the job has been accepted and a machine must be selected from the ones that the molder has.
2. The list of activities and job functions is a general list and every company has its own organizational structure. Therefore, every company must generate a customized matrix. A meeting or updates at the end of every stage should be mandatory. Although all job functions may not have a direct role in every stage, their decisions will be based on the information provided to them from the previous stages. Therefore, the status and decisions made regarding the project must be communicated to the whole group.

The molding business is getting extremely competitive in terms of cost and lead times. The time lines from conceptualizing the project to having an actual molded part are shrinking and it is expected that acceptable parts are produced on the first molding attempt. Only companies that can achieve this are likely to survive in the more and more competitive market. Concurrent engineering should be practiced because it provides the extra set of eyes to 'Look out of the Box' and raise potential concerns. These are meetings and times well spent. Regular reviews must be done, especially if there is a change in design, material, time line, etc., and must be communicated to everyone. The final product, good or bad, is a result of the involvement of the whole team.

Appendix A
Materials Data Sheet

List of processing related data for commonly injection molded materials – SI system. (Courtesy www.ides.com)

Note: These are guidelines only – Individual datasheets must be obtained from the material suppliers or from www.ides.com

Long name	Short name	Specific gravity	Mold shrinkage, flow %
Polycarbonate + PBT	PC+PBT	1.10 – 32.1	0.5 – 1.0
Polycarbonate + PET	PC+PET	1.20 – 1.22	0.5 – 0.9
Polybutylene terephthalate	PBT	1.00 – 33.6	0.1 – 0.5
Polyethylene terephthalate	PET	1.32 – 1.41	0.3 – 1.8
Polyethylene terephthalate glycol comonomer	PETG	1.25 – 1.28	0.3 – 0.5
Polyether imide	PEI	1.26 – 1.36	0.1 – 0.2
Polyetheretherke-ne	PEEK	1.25 – 1.40	0.1 – 1.7
Ethylene vinyl acetate copolymer	EVA	0.929 – 0.962	1.2 – 1.5
Polyethylene, high density	HDPE	0.932 – 0.988	0.7 – 3.0
Polyethylene, low density	LDPE	0.893 – 0.955	1.3 – 3.1
Polyethylene, linear low density	LLDPE	0.918 – 0.948	1.5 – 2.0
Polyethylene, ultra high molecular weight	UHMWPE	0.920 – 0.947	0.6 – 3.0
Polylactic acid	PLA	1.23 – 1.26	0.3 – 1.1
Polyphenylene ether	PPE	1.04 – 1.10	0.6
Polyphenylene sulfide	PPS	1.26 – 1.75	0.1 – 0.3
Polypropylene homopolymer	PP Homopolymer	0.901 – 0.950	1.2 – 1.8
Polystyrene, general purpose	PS (GPPS)	1.03 – 1.06	0.4 – 0.6
Polystyrene, high impact	PS (HIPS)	1.03 – 1.06	0.4 – 0.6
Syndiotactic polystyrene	SPS	1.01 – 1.44	0.3 – 2.0
Polyether sulfone	PES	1.37 – 1.38	0.6 – 1.4
Polysulfone	PSU	1.24 – 1.25	0.6 – 1.0
Polyvinyl chloride, chlorinated	CPVC	1.47 – 1.52	0.6
Polyvinyl chloride, flexible	PVC, Flexible	1.14 – 1.45	0.9 – 2.1
Polyvinyl chloride, rigid	PVC, Rigid	0.779 – 1.47	0.3 – 0.4
Polyvinyl chloride, semi-rigid	PVC, Semi-Rigid	1.30 – 1.58	1.1

Drying temperature (°C)		Drying time (h)	Suggested max moisture (%)	Processing (melt) temp (°C)		Mold temperature (°C)	
min	max			min	max	min	max
94	121	2.0 – 5.0	0.020 – 0.022	257	272	62	105
97	118	2.0 – 8.0	0.019 – 0.020	267	271	79	81
113	132	3.0 – 6.0	0.020 – 0.043	234	265	58	92
120	180	4.0 – 5.5	0.0030 – 0.20	256	285	15	130
65	75	3.0 – 9.0	0.05	218	260	27	40
134	151	5.0 – 5.5	0.020 – 0.021	373	374	148	151
135	150	3.0 – 4.0	0.1	374	384	149	192
60	61	7.0 – 8.0	NA	98	230	20	180
79	80	1.0	NA	180	251	10	46
79	82	1.0	NA	164	222	15	43
90		1.0	NA	180	240	18	35
NA	NA	1.0	NA	285		45	–
49	51	3.5 – 4.0	0.010 – 0.032	199	240	20	105
70	125	3	0.02	260	315	74	90
134	150	4.0 – 6.0	0.015 – 0.20	313	323	121	156
75	85	1.0 – 3.0	0.050 – 0.20	202	249	2	50
69	82	1.5 – 2.0	0.02	214	248	30	60
70	78	1.5 – 2.0	0.1	208	236	29	61
80		3.5	NA	310		70	–
134	177	2.5 – 6.0	0.020 – 0.050	355	366	134	160
134	149	3.0 – 4.0	0.020 – 0.10	352	366	121	151
NA	NA	NA	NA	203	204	NA	NA
NA	NA	NA	NA	165	200	24	30
66		3	NA	186	206	32	32
NA	NA	NA	NA	188	194	NA	NA

Long name	Short name	Specific gravity	Mold shrinkage, flow %
Styrene acrylonitrile	SAN	1.04 – 28.1	0.3 – 0.5
Styrene butadiene block copolymer	SBC	0.850 – 1.03	0.5 – 2.4
Styrene butadiene styrene block copolymer	SBS	0.922 – 1.05	0.4 – 1.4

Drying temperature (°C)		Drying time (h)	Suggested max moisture (%)	Processing (melt) temp (°C)		Mold temperature (°C)	
min	max			min	max	min	max
77	80	2.0 – 4.0	0.020 – 0.20	204	251	49	65
52	77	0.5 – 3.0	NA	184	250	39	50
52	52	0.0 – 3.0	NA	155	232	23	45

Appendix B
Conversion Tables for Commonly Used Process Parameters

Speed

1 mm/s = 0.0394 in/s
1 in/s = 25.4 mm/s

Pressure

1 psi = 0.0069 MPa
1 psi = 0.0690 bar

1 bar = 14.50 psi
1 bar = 0.1 MPa

1 MPa = 145.04 psi
1 MPa = 10 bar

Temperature

°Fahrenheit = (1.8 × °Centigrade) + 32
Example: 50 °C = (1.8 × 50) + 32 = 122 °F

°Centigrade = (°Fahrenheit – 32) / 1.8
Example: 200 °F = (200–32)/1.8 = 93.3 °C

Weight

1 ounce (oz)= 28.35 g
1 gram (g) = 0.035 oz

Tonnage

1 kN = 0.11 US tons
1 kN = 0.1 metric ton

1 US ton = 9.09 kN
1 US ton = 0.909 metric tons

1 metric ton = 10 kN
1 metric ton = 1.1 US tons

Appendix C
Water Flow Tables

Minimum Water Flow (gal/min) Required to Achieve Turbulent Flow

Water temperature (°F)	Pipe diameter in inches				
	¼"	3/8"	1/5"	¾"	1"
	(0.25 in)	(0.375 in)	(0.5 in)	(0.75 in)	(1 inch)
50	0.41	0.62	0.82	1.23	1.64
60	0.35	0.53	0.71	1.06	1.42
70	0.31	0.46	0.62	0.93	1.23
80	0.27	0.41	0.54	0.81	1.09
90	0.24	0.36	0.48	0.72	0.96
100	0.22	0.32	0.43	0.65	0.86
125	0.17	0.25	0.33	0.50	0.67
150	0.13	0.20	0.27	0.40	0.54
175	0.11	0.17	0.22	0.34	0.45
200	0.09	0.14	0.19	0.28	0.38

Note: This table should be used for water only. The presence of additives, such as rust preventives, will alter the viscosity and therefore the flow rates. It is always better to use more than the recommended flow rates for this reason.

Minimum Water Flow (l/min) Required to Achieve Turbulent Flow

Water temperature (°C)	**Pipe diameter in mm**				
	8	**10**	**15**	**20**	**25**
10	1.96	2.45	3.68	4.90	6.13
15	1.71	2.14	3.21	4.29	5.36
20	1.51	1.89	2.83	3.78	4.72
25	1.34	1.68	2.52	3.36	4.20
30	1.20	1.50	2.26	3.01	3.76
35	1.08	1.36	2.03	2.71	3.39
40	0.98	1.23	1.84	2.46	3.07
45	0.90	1.12	1.68	2.24	2.80
55	0.76	0.94	1.42	1.89	2.36
65	0.65	0.81	1.22	1.62	2.03
80	0.53	0.66	0.99	1.32	1.66
95	0.44	0.56	0.83	1.11	1.39

Note: This table should be used for water only. The presence of additives, such as rust preventives, will alter the viscosity and therefore the flow rates. It is always better to use more than the recommended flow rates for this reason.

Appendix D
Part Design Checklist

Following are examples of some of the important questions that must be asked during a part design review:

Part Design Checklist

- Is the gate location acceptable for fill?
- Is the number of gates sufficient for part fill?
- Is the gate location acceptable for cosmetics?
- Is the gate location acceptable for warpage?
- Can the thick areas be cored out?
- Is processing information about the plastic material available?
- Is the shrink factor available?
- Is the engraving in an acceptable position?
- Is the texture acceptable?
- Is the part designed to stay on the B-side of the mold?
- Is sufficient draft provided on the part for ejection?

Appendix E
Mold Design Checklist

Following are examples of some of the important questions that must be asked during a mold design review:

1. Mold Cooling:

a. Is the mold cooling acceptable?
b. Does the mold require waterlines or oil lines?
c. What is the size of the waterlines?
d. Are the waterlines recessed?

2. Mold Construction:

a. Is the 'TOP' of the mold acceptable for part ejection and removal?
b. Are eyebolt holes provided on the 'TOP' of the mold?
c. Will the mold fit between the tie bars?
d. Are the clamp slots/holes acceptable?
e. Do the waterlines and tie bars interfere with each other?
f. Are parting line interlocks necessary?
g. Are pry bar slots necessary?
h. Is the mold opening sequence acceptable?
i. Are insulator plates necessary?
j. Are any spares necessary?
k. Are the machine tie bars shown on the drawings?

3. Ejection:

a. Is the number of ejector pins sufficient?
b. Are the sizes of ejector pins sufficient?
c. Does the ejector pattern on the mold match the ejector pattern on the machine?
d. Is the ejection distance sufficient for the part to get ejected out of the mold?
e. Are return springs necessary?
f. Is an early ejector return system necessary?
g. Are support pillars necessary?
h. Is guided ejection necessary?
i. Does the path of any of the ejector pins and the slides interfere with each other?

4. Venting:

a. Are vents identified on the mold?
b. Is the last point to fill vented?
c. Are all the corners vented?
d. Are bosses, ribs and tabs vented?
e. Are deep pockets vented?
f. Are the runners vented?

5. Gates and Runners:

a. Is the gate size acceptable?
b. Will the gate leave a gate vestige?
c. Is the runner size acceptable?
d. Is the runner balanced?
e. Will the runners stay on the B-side?
f. In case of a three-plate mold will the runner drop freely?
g. Is there a sprue puller?
h. Do any of the sucker pins interfere with the plastic flow?

Appendix F
Mold Qualification Checklist:

a. Are the gates acceptable?
b. Are the parts dimensionally acceptable?
c. Has the cooling study been done?
d. Has the gate seal test been done?
e. Has the viscosity test been performed?
f. Is the cavity balance acceptable?
g. Is the cycle time acceptable?
h. Is the ejection acceptable?
i. Is the part fill acceptable?
j. Is the process window acceptable?
k. Is the runner acceptable?
l. Is the venting acceptable?
m. Is the water flow through the mold acceptable?
n. Was the mold safe to be hung in the press?

Appendix G
Regrind Tables – Percentage of regrind in total shot.

(The runner and part will always have the same percentage of regrind.)

Table 1 Part Weight = 95 % Runner Weight = 5 %

Regrind Generation (g)	Pass (p)				
	1	2	3	4	5
0	100	95.00	95.00	95.00	95.00
1	-	5.00	4.75	4.75	4.75
2	-	-	0.25	0.24	0.24
3	-	-	-	0.01	0.01
4	-	-	-	-	0.00

Table 2 Part Weight = 90 % Runner Weight = 10 %

Regrind Generation (g)	Pass (p)				
	1	2	3	4	5
0	100	90.00	90.00	90.00	90.00
1	-	10.00	9.00	9.00	9.00
2	-	-	1.00	0.90	0.90
3	-	-	-	0.10	0.09
4	-	-	-	-	0.01

Table 3 Part Weight = 80 % Runner Weight = 20 %

Regrind Generation (g)	Pass (p)				
	1	2	3	4	5
0	100	80.00	80.00	80.00	80.00
1	-	20.00	16.00	16.00	16.00
2	-	-	4.00	3.20	3.20
3	-	-	-	0.80	0.64
4	-	-	-	-	0.16

Table 4 Part Weight = 75 % Runner Weight = 25 %

Regrind Generation (g)	Pass (p)				
	1	2	3	4	5
0	100	75.00	75.00	75.00	75.00
1	-	25.00	18.75	18.75	18.75
2	-	-	6.25	4.69	4.69
3	-	-	-	1.56	1.17
4	-	-	-	-	0.39

Table 5 Part Weight = 50 % Runner Weight = 50 %

Regrind Generation (g)	Pass (p)				
	1	2	3	4	5
0	100	50.00	50.00	50.00	50.00
1	-	50.00	25.00	25.00	25.00
2	-	-	25.00	12.50	12.50
3	-	-	-	12.50	6.25
4	-	-	-	-	6.25

Table 6 Part Weight = 25 % Runner Weight = 75 %

Regrind Generation (g)	Pass (p)				
	1	2	3	4	5
0	100	25.00	25.00	25.00	25.00
1	-	75.00	18.75	18.75	18.75
2	-	-	56.25	14.06	14.06
3	-	-	-	42.19	10.55
4	-	-	-	-	31.64

Subject Index